STORMWATER, WATERSHED, AND RECEIVING WATER QUALITY MODELING

WEF Special Publication

2020

Water Environment Federation
601 Wythe Street
Alexandria, VA 22314-1994 USA
www.wef.org

Stormwater, Watershed, and Receiving Water Quality Modeling

ISBN: 978-1-57278-359-1

About WEF

The Water Environment Federation (WEF) is a not-for-profit technical and educational organization of 35,000 individual members and 75 affiliated Member Associations representing water quality professionals around the world. Since 1928, WEF and its members have protected public health and the environment. As a global water sector leader, our mission is to connect water professionals; enrich the expertise of water professionals; increase the awareness of the impact and value of water; and provide a platform for water sector innovation. To learn more, visit www.wef.org.

Prepared by the **Stormwater, Watershed, and Receiving Water Quality Modeling** Task Force of the **Water Environment Federation**

Caroline Burger, *Chair*
Nitin Katiyar, P.E., *Vice-Chair*
Steven Wolosoff, BCES, *Vice-Chair*

Ayman Alafifi, Ph.D.
Michael Beezhold, CPM
Arturo Burbano, Ph.D., P.E., BCEE
Rajat Chakraborti, Ph.D.
Chein-Chi Chang, Ph.D., P.E.
Zachary Eichenwald, P.E.
Sara Ferrance
Dan Figola
Kim Grove
Richard Haimann
Mahmudul Hasan
Aiza F. Jose Sanchez, Ph.D., P.E.
Jagjit Kaur, Ph.D.
Sudhir Kshirsagar, Ph.D., P.E.
Rishab Mahajan
Henry Manguerra, Ph.D., P.E.
Dmitrijs Obolevics, IEng MICE
Fernando Pasquel
Sabu Paul, Ph.D., P.E., PMP
Christopher Ranck, P.E., BCEE, D.WRE, ENV SP
Debabrata Sahoo, Ph.D., P.E., P.H.
Michael Schmidt, P.E.
John Schooler, Jr., P.E.
Chelsea Stroble, M. Eng.
Anil Tangirala
Shelby M. Taylor, M.S.
Saša Tomić, Ph.D., P.E., BCEE, D.WRE
Gian Villarreal, P.E., CFM, CPSWQ
Richard Wagner, P.E.
Harry Zhang, Ph.D., P.E.

Under the Direction of the **Stormwater Subcommittee** of the **Technical Practice Committee.**

2020

Water Environment Federation
601 Wythe Street
Alexandria, VA 22314-1994 USA
http://www.wef.org

Special Publications of the Water Environment Federation

The WEF Technical Practice Committee (formerly the Committee on Sewage and Industrial Wastes Practice of the Federation of Sewage and Industrial Wastes Associations) was created by the Federation Board of Control on October 11, 1941. The primary function of the Committee is to originate and produce, through appropriate subcommittees, special publications dealing with technical aspects of the broad interests of the Federation. These publications are intended to provide background information through a review of technical practices and detailed procedures that research and experience have shown to be functional and practical.

Water Environment Federation Technical Practice Committee Control Group

Contents

List of Figures

List of Tables

Preface

This publication fills a need for a comprehensive and up-to-date stormwater quality and modeling manual for the industry. It focuses on water quality models—models that predict volumes and loads from the land surface, both urban and rural, and then route the volume and pollutant loading through the receiving waters. It summarizes the history of water quality modeling, provides an overview of currently available tools, and outlines criteria for selecting the appropriate model for one's current project or task. This publication also features a discussion on the future of the water quality modeling industry.

This publication was produced under the direction of Caroline Burger, *Chair*, Nitin Katiyar, *Vice-Chair*, and Steven Wolosoff, *Vice-Chair*.

Authors' and reviewers' efforts were supported by the following organizations:

AECOM, Madison, Wisconsin, USA
Arcadis U.S., Inc., Indianapolis, Indiana, USA
Arup, New York, New York, USA
Atkins, a member of the SNC-Lavalin Group, Alexandria, Virginia, USA
Atkins, a member of the SNC-Lavalin Group, Los Angeles, California, USA
Atkins Global, Dallas, Texas, USA
BaySaver Technologies, Hilliard, Ohio, USA
Brown and Caldwell, Madison, Wisconsin, USA
Brown and Caldwell, Seattle, Washington, USA
Carollo Engineers, Inc., Arlington, Virginia, USA
CDM Smith, Inc., Boston, Massachusetts, USA
CDM Smith, Inc., Cambridge, Massachusetts, USA
CDM Smith, Inc., Maitland, Florida, USA
Department of Public Works, Baltimore City, Maryland, USA
District of Columbia Water and Sewer Authority, Washington, D.C., USA
DPW Plans Review and Inspections, Baltimore City, Maryland, USA
Geosyntec Consultants, Oak Brook, Illinois, USA
Global Foundation for Environmental QULA, Huntington Beach, California, USA
Global Quality Corp, Covington, Kentucky, USA

GHD Services Inc., Miami Lakes, Florida, USA
HDR, Mahwah, New Jersey, USA
HDR, New York, New York, USA
Jacobs Engineering Group Inc., Los Angeles, California, USA
Kansas Department of Health and Environment, Topeka, Kansas, USA
MS Consultants, Inc., Columbus, Ohio, USA
Seagull PME, Oceanside, California, USA
The George Washington University, Washington, D.C., USA
The Water Research Foundation, Alexandria, Virginia, USA
Woolpert, Inc., Chesapeake, Virginia, USA
Woolpert Inc, Columbia, South Carolina, USA

1

Introduction

Aiza F. Jose Sanchez, Ph.D., P.E. and Sara Ferrance

1.0 IMPETUS FOR THE MANUAL

The idea for this manual was first brought forth by the Water Environment Federation (WEF) Stormwater and Watershed Committees in December 2014. At that time, many members of the committees were still referencing stormwater manuals that were approximately 20 years old. Although the basic theory presented in the manuals remains applicable to today's stormwater industry, many of the water quality models described in the publications are no longer used or have undergone significant redevelopment. Indeed, advances in technology and the state of practice, along with more stringent permitting requirements, call for the improved use of geospatial and remote-sensing data and better integration of quality and quantity models to seamlessly exchange data and outputs between different decision support tools. For these reasons, the committee members identified a need for a comprehensive and up-to-date stormwater quality and modeling manual for the industry.

As a starting point for this effort, the WEF Stormwater and Watershed Committees evaluated the 1997 U.S. Environmental Protection Agency (U.S. EPA) *Compendium of Tools for Watershed Assessment and TMDL Development* (U.S. EPA, 1997). The compendium is a comprehensive and readily accessible tool that has remained one of the most-referenced sources by modelers for over 20 years. The compendium ultimately served as a guide for the development of this manual.

2.0 CONTENTS

This manual takes a "past, present, and future" look at the stormwater quality modeling industry. Chapters 2 and 3 summarize the history of stormwater modeling, early modeling efforts, and the regulatory framework affecting the need for modeling. Chapters 4 and 5 discuss how current models are used in watershed, urban stormwater, and receiving water modeling, addressing the models' applicability and practicality. Chapter 6 describes the main drivers for future stormwater quality modeling applications. Chapter 7 compiles model-specific information for each of the models discussed in this manual. The following paragraphs contain more detailed descriptions of the information presented in each chapter.

Chapter 2 presents a brief overview of the history and evolution of stormwater modeling practices. Starting with a description of pollution and its effects, it describes the development of the regulatory framework, early developments in modeling efforts, and the relationship between stormwater quantity and quality. The chapter further describes the evolution of pollutant quantification and prediction derived from hydraulic equations and chemical reactions. It also discusses initial modeling efforts, presenting the limitations in data processing and collection and the roles that scale, complexity, and regional variability play.

Chapter 3 presents the evolution of models used for stormwater quality, identifying the current drivers—including advances in technology, research, and the regulatory framework—associated with water protection. Technology is discussed in detail because of the significance of major advances in computers (both software and hardware), geographic information systems (GIS) and graphical user interfaces (GUIs), remote sensing data acquisition, and communication (i.e., the Internet and online database managing capabilities). Other drivers identified include an enhanced environmental awareness, advances in water quality planning and pollution control, advances in stormwater management incorporating low impact development (LID) and green infrastructure, and the integrated approach for the management of surface water and groundwater interactions.

Future drivers for the evolution of the models discussed include future stormwater management regulations, the changing nature of pollution problems, emerging contaminants and pathogens, the increasing need for optimization techniques to develop economical solutions, the One Water holistic approach for the management of water systems, and consideration of extreme weather events affecting stormwater quantity and quality.

Chapter 4 describes the current models used to simulate water quality that are typically categorized into two groups: watershed models and instream/receiving water models. Watershed models simulate the processes of various land uses and practices, handling both urban and rural environments, with varying degrees of modeling capabilities with respect to scale, simulation time step, and types of land uses, pollutants, and diversity of best management practices (BMPs). Instream or receiving water modeling tools emphasize hydrology and water of conveyance systems such as rivers, reservoirs, lakes, and estuaries. For practical application, Chapter 4 groups the different models into the following categories: watershed quality models, urban stormwater quality models, receiving water quality models, integrated modeling systems, and water quality compliance models. Watershed quality models (for the purpose of classification in this chapter) are those that handle both rural or agricultural environments, whereas urban stormwater modeling tools simulate urban landscape and stormwater networks. The integrated models consist of a combination of individual models or tools into a common platform. This common platform allows the models to become linked so that the output from one model can be accessed as the input for another model in a way the models can be run either seamlessly or in a sequence. Water quality compliance models are those models that are currently used for meeting regulatory stormwater compliance.

The water quality models included in this manual were selected as follows. As stated previously, the 1997 U.S. EPA *Compendium of Tools for Watershed Assessment and TMDL Development* (U.S. EPA, 1997) was used as a starting point. The authors of this book analyzed different publicly available stormwater quality models and then categorized the models into "legacy" or "current" according to their documented continuous application: state-of-the-art and/or state-of-practice. Current publicly available models are the focus of this manual. The model fact sheets in Chapter 7 list the widely known current proprietary and commercial versions of a current model. Finally, legacy models were not selected for discussion in this manual.

Chapter 5 presents practical applications for the modeler relating to the model selection process, which is mainly determined by modeling objectives, spatial and temporal scales, and other model requirements and capabilities. The chapter presents comparison tables for both land-based and receiving water quality simulation models and their capacity to model different spatial and temporal scenarios, constituents of interest, capability to model additional stormwater management measures (e.g., detention basin, street sweeping, fertilizer and pesticides application, livestock fencing, etc.), and available technical support from the models' developers. Key considerations for model selection analyzed in Chapter 5 include the level of results expected (screening vs. detailed), spatial extent (site vs. watershed), land cover (agricultural, urban, rural, or mix urban and rural), temporal scale (event, annual, continuous, combination of those), type of waterbody (stream, lake, estuary, or combination of those), dimensionality (1-, 2-, 3-dimensional models), and spatial extent/segmentation.

The selection process for appropriate modeling tools is also based on the type of constituents and process that need to be modeled. Constituents analyzed in land-based models include water temperature, dissolved oxygen, biochemical oxygen demand, sediment, nitrogen, phosphorous, bacteria, pesticides, metals, alkalinity, pH, and others. Constituents available for receiving water models include the ones listed for land-based models, plus phytoplankton (free-floating algae in receiving waters, typically measured based on chlorophyll-a content); periphyton (typically forming on rocky bottoms and where there is sufficient light penetration); macrophytes (attached plants, primarily on the fringes of a receiving water, that are rooted underwater but extend above the water surface); zooplankton (organisms that float in the water and feed on the phytoplankton); total dissolved solids; salinity; and other constituents. Processes modeled for land-based models include runoff and groundwater flow and quality. Processes modeled for receiving water models include sediment transport and reaeration.

Chapter 6 presents the future of stormwater quality modeling based on the drivers identified in Chapter 3 and two types of model advancement trajectories: (a) anticipated "normal" advancement trajectory resulting from expected developments in technology and data and (b) desired trajectory as driven by current and predicted modeling and functional needs (e.g., discipline gaps related to scientific rigor, ecological response, ecosystems services and socio-economic impact, BMP integration, model integration). In terms of advances in data, technology, and science, Chapter 6 discusses important drivers for the evolution of modeling software, including (a) further advances in the use of GUIs that allow for the handling of increasingly complex model applications into intuitive solutions for a broader range of users, from the experienced modeler to the novice, and are easily modularized depending on the complexity and size of the system; (b) seamless incorporation of large sets of data and the ability of the models to automatically download, check for quality issues, and format data from multiple sources into model input and then generate visual reports of model results that can be shared with decision-makers; (c) improvement in the use of data collected by sensor technology such as light detection and ranging (LiDAR) and multispectral imagery; (d) the need to develop a common framework that couples and integrates models with different capabilities or different model components; and (e) the need for models to be documented and packaged with data and be made readily accessible for other researchers and modelers to use.

In terms of model evolution in response to business needs and anticipated discipline gaps, advances in scientific rigor are expected to improve components of some of the less well-understood modeling processes, including erosion; sediment transport and deposition; fate and transport of selected water constituents; ecological response to hydrologic, hydraulic, and water quality changes; and integration of ecosystem services and socio-economic impact assessments in the modeling efforts. Modifications may be also expected in the strategies used to integrate BMPs into the models. Modeling of BMPs may evolve from representation as point sinks (using performance efficiencies) or as explicit elements of a landscape/conveyance/transport system (where detention, infiltration, evaporation, and others are simulated) to solutions in which BMPs are modeled more mechanistically within the water system, allowing comparison and contrast of alternative BMP scenarios. Evolution of water quality models may also allow the integration of climate change projections and the incorporation of integrated water resources management (adding wastewater and water supply components). Models may also feature dashboards that allow modelers, decision-makers, and stakeholders to mine, synthesize, and visualize different alternative scenarios with single or multiple objectives; allow for real-time operation, maintenance, and disaster mitigation; and facilitate distributed and participatory modeling with the use of advances in cloud computing, data connectivity, and web-based user interfaces.

Chapter 6 also provides recommendations moving forward for the stormwater quality modeling industry utilizing an organized and systematic collaborative approach between stormwater professionals to facilitate and support the effective evolution of this practice.

Chapter 7 presents a summary of each model discussed in this manual. Chapter 7 contains the following information:

1. Model availability—source of the model, history of the model, online availability, online user group if available, link to the model if available, key words, versions, support, and cost;
2. Types of modeling and potential application areas—simulation type, nature of the modeling, and recommended extents (watersheds, sites); general application of the model, such as agricultural, urban, semi-urban, and so on (provides references for different types of applications);
3. Pollutants—types of pollutants that the stormwater model can simulate;
4. Model components, techniques, and processes—rainfall-flow process, sediment, and nutrient transport;

5. Input—data requirements;
6. Simulation outputs—graphical output, numerical outputs, time-step output, and so on;
7. Model limitations; and
8. References.

3.0 WHAT IS AND IS NOT INCLUDED

This manual focuses on stormwater quality models—models that generate surface runoff volume and pollutant loads and/or route the volume and pollutant loads through the receiving waterbodies. Because stormwater quality concerns originate from the first flush of stormwater runoff, the manual emphasizes discussion of those flows that mainly contribute to the water quality of a waterbody, focusing less on larger storm events, such as peak flow events. For the benefit of the user of this manual, readily and publicly available references to peak flow routing and its effect on peak loading appear at the end of this chapter.

This manual is a practical tool that presents a broad compilation of existing models, their capabilities, and a roadmap for users to assist in the selection of such models for diverse applications. However, the manual does not address the fate and transport theory or the mathematical methodology behind those models. The audience for this manual are industry users with an underlying knowledge and understanding of the theory behind water quantity and quality modeling.

This manual focuses on presenting state-of-the-art stormwater modeling in the United States. Although the concepts would be equally applicable elsewhere internationally, having a perspective of application of those models elsewhere is key for their practical use. Referencing international studies and analysis may be used by the practitioner to generate the perspective necessary to extrapolate this manual application to the international arena. Based on limited research of readily available literature on international stormwater quality modeling applications, recent stormwater modeling efforts have been documented in France, Taiwan, Australia, the Netherlands, Estonia, India, Austria, East Asia, and Denmark. Some of the models reportedly used in these studies include some of those commonly used in the United States such as SWMM, MIKE, Streeter-Phelps, QUAL, WASP, QUASAR, BASIN, and EFDC models. Some references report the use of models such as GLUE and a multi-model Bayesian approach.

Because of its described limitations, this manual can be complemented using other readily available stormwater modeling manuals and other recent literature, including the 2017 American Society of Civil Engineers

(ASCE) article in the *Journal of Hydrologic Engineering,* "Special Collection on Total Maximum Daily Load Analysis and Modeling: Assessment and Advancement". The recommended reading section of this chapter lists some of those manuals and publications.

4.0 REFERENCES

U.S. Environmental Protection Agency. (1997) *Compendium of tools for watershed assessment and TMDL development.* Washington, D.C.: U.S. Environmental Protection Agency.

5.0 SUGGESTED READING

Akan, A. O., & Houghtalen, R. J. (2003). *Urban hydrology, hydraulics, and stormwater quality: Engineering application and computer modeling.* Hoboken, NJ: Wiley.

American Society of Civil Engineers. TMDL Analysis and Modeling Task Committee (1st ed.). (2017). *Total maximum daily load analysis and modeling: Assessment of the practice.* Reston, VA: American Society of Civil Engineers.

Bertrand-Krajewski, J. L. (2007). Stormwater pollutant loads modelling: epistemological aspects and case studies on the influence of field data sets on calibration and verification. *Water Science and Technology, 55*(4), 1–17.

Borah, D. K., Ahmadisharaf, E., Padmanabhan, G., Imen, S., & Mohamoud Y. M. (2019). Watershed models for development and implementation of total maximum daily loads. *Journal of Hydrologic Engineering,* 4(1).

Camacho, R. A., Zhang, Z., & Chao, X. (2019). Receiving water quality models for TMDL development and implementation. *Journal of Hydrologic Engineering,* 24(2).

Chapra, S. C. (2008). *Surface Water-Quality Modeling.* Long Grove, IL: Waveland Press, Inc.

Chiang, Y.-M., Chang, L.-C., Tsai, M.-J., Wang, Y.-F., & Chang, F.-J. (2010). Dynamic neural networks for real-time water level predictions of sewerage systems-covering gauged and ungauged sites. *Hydrology and Earth System Sciences, 12,* 1309–1319.

Dotto, C. B. S., Kleidorfer, M., Deletic, A., Rauch, W., Mccarthy, D.T., & Fletcher, T. D. (2011). Performance and sensitivity analysis of stormwater models using a Bayesian approach and long-term high-resolution data. *Environmental Modelling & Software,* 26(10), 1225–1239.

Ji, Z.-G. (2017). *Hydrodynamics and water quality: Modeling rivers, lakes, and estuaries*. Hoboken, NJ: Wiley & Sons.

Krešić, N. (2007). *Hydrogeology and groundwater modeling*. Boca Raton, FL: CRC Press.

Kroeze, C., Gabbert, S., Hofstra, N., Koelmans, A. A., Li, A., Löhr, A., ... & Wijnen, J. V. (2016). Global modelling of surface water quality: A multi-pollutant approach. *Current Opinion in Environmental Sustainability, 23,* 35–45.

Maharjan, B., Pachel, K., & Loigu, E. (2017). Modelling stormwater runoff, quality, and pollutant loads in a large urban catchment. *Proceedings of the Estonian Academy of Sciences, 66*(3), 225.

McCutcheon, S. C. (1989). *Transport and surface exchange in rivers*. Boca Raton, FL: CRC Press.

Meals, D. W., Richards, R. P., & Dressing, S. A. (2013). Pollutant load estimation for water quality monitoring projects. *Tech Notes 8*, April 2013. Developed for the U.S. Environmental Protection Agency by Tetra Tech, Inc.: Fairfax, VA.

Minnesota Pollution Control Agency. (2017). Available stormwater models and selecting a model. *Minnesota Stormwater Manual*. Retrieved January 3, 2017, from https://stormwater.pca.state.mn.us/index.php/Available_stormwater_models_and_selecting_a_model

Mohamoud, Y., & Zhang, H. (2019). Applications of linked and nonlinked complex models for TMDL development: Approaches and challenges. *Journal of Hydrologic Engineering*, *24*(1).

National Research Council. (2009). *Urban Stormwater Management in the United States*. Washington, D.C.: National Academies Press.

Orlob, G. T. (1979). Mathematical modelling and simulation of water quality: A survey of the state-of-the-art. *Hydrological Sciences Journal, 24*(2), 151–156.

Palmer, M. D. (2001). *Water quality modeling: A guide to effective practice.* Washington D.C.: The World Bank.

Sumita, N., & Kaur, B. S. (2017). Water quality models: A review. *International Journal of Research – Granthaalayah, 5*(1), 395–398.

Thomann, R. V., & Mueller, J. A. (1993). *Principles of surface water quality modeling and control*. New York, NY: Harper Collins Publishers.

U.S. Army Corps of Engineers. (1979). *Flood hydrograph and peak flow frequency analysis* (TP-62). Davis, CA: Hydraulic Engineering Center.

U.S. Army Corps of Engineers. (1996). *Modeling water resource systems for water quality management* (TP-154). Davis, CA: Hydraulic Engineering Center.

U.S. Environmental Protection Agency. (2009). *Technical guidance on implementing the stormwater runoff requirements for federal projects under Section 438 of the Energy Independence and Security Act* (EPA 841-B-09-001). Washington, D.C.: U.S. Environmental Protection Agency, Office of Water (4503T).

U.S. Environmental Protection Agency. (2012). *Compendium of tools for watershed assessment and TMDL development*. Washington, D.C.: Bibliogov Project.

U.S. Environmental Protection Agency. (2013). *Rate, constants, and kinetics formulations in surface water quality modeling*. Washington, D.C: Bibliogov Project. Retrieved from https://nepis.epa.gov/Exe/ZyPURL.cgi?Dockey=9100R3IW.TXT

Vezzaro, L., Ledin, A., & Mikkelsen, P. S. (2012). Integrated modelling of priority pollutants in stormwater systems. *Physics and Chemistry of the Earth*, 42–44, 42–51.

Yeo, I.-Y., Gordon, S. I., & Guldmann, J.-M. (2004). Optimizing patterns of land use to reduce peak runoff flow and nonpoint source pollution with an integrated hydrological and land use model. *Earth Interactions*, *8*, 1–20.

2

History of Stormwater, Watershed, and Receiving Water Quality Modeling

Chelsea Stroble, M. Eng. and Shelby M. Taylor, M.S.

1.0 THE EFFECTS OF POLLUTION

Pollution is one of the main reasons for water quality modeling. By examining the observed effects of pollution on public health, the observed effects on aquatic life, the relationship of pollution to weather conditions, and the relationship of pollution to human activities, a connection between the effects of pollution and water quality modeling can be made.

The Department of Energy and Environment (DOEE), a District of Columbia agency, defines stormwater runoff effects as follows (DOEE, n.d.):

- Flooding—damage to public and private property;
- Eroded streambanks—sediment clogs waterways, fills lakes, reservoirs, and kills fish and aquatic animals;
- Widened stream channels—loss of valuable property;
- Aesthetics—dirty water, trash and debris, foul odors;
- Fish and aquatic life—impaired and destroyed;
- Impaired recreational uses—swimming, fishing, boating;
- Threatens public health—contamination of drinking water, fish/shellfish;
- Threatens public safety—drownings occur in flood waters;
- Economic effects—impairments to fisheries, shellfish, tourism, recreation-related businesses; and
- Increased cost of water and wastewater treatment—stormwater pollution increases raw water treatment costs and reduces the assimilative capacity of waterbodies.

1.1 Observed Effects on Public Health

As urban areas increased in size and number, channeling waterways was required for development. These areas merged residential and business structures, disrupting the natural flow of waterways. This caused unforeseen waterborne diseases such as cholera and typhoid. Therefore, requirements to separate and maintain the movement of drinking water from surface water were rendered to prevent contamination. In 1908, Jersey City, New Jersey, became the first city to treat their drinking water with chlorine. The first federal regulations for drinking water were implemented in 1919 in the form of standards for bacteriological quality. Improvements implemented in 1962 increased the standards to include the regulation of 28 substances. Finally, in 1974, the Safe Water Drinking Act was created and adopted by all 50 states, establishing standards and guidelines for public water systems (U.S. EPA, 2000).

1.2 Human Effect and Cause of Stormwater Management

Urban areas are created mostly from impervious surfaces such as sidewalks, roadways, parking lots, building structures, and construction sites. With these surfaces come human-made substances such as oil, gas, automotive fluids, hydrocarbons, chemicals, fertilizers, pesticides, sediments, hazardous materials, and unwanted nutrients. These substances are physical, chemical, and biological contaminates. During storms, these contaminates are transported by stormwater management systems to natural bodies of water.

The influx of sediments and substances to these waterways and groundwater cause harmful and poisonous effects to the environment. Impervious surfaces prevent soil from filtering contaminants and increase the speed of surface runoff. Characteristics of runoff that help shape stormwater modeling are defined in the following three categories:

- Pollutant transport
- Runoff velocity
- Point-of-runoff reduction

The area in which runoff is confined greatly affects its velocity. Smaller spaces will cause runoff to flow rapidly and at great depth (e.g., runoff spilling onto a sidewalk before a storm drain). Rapid runoff is a contributing factor to erosion of many human-made and natural areas. An example of this is the disruption of riverbeds in which sediment deposits collect in pockets between rocks, stagnating water and creating breeding areas for pathogenic bacteria and viruses. An effect on aquatic life is seen in stream studies when the depletion of specific macro invertebrate species becomes apparent.

Non-point-source pollution such as stormwater runoff is one of the U.S. Environmental Protection Agency's (U.S. EPA) most important sources of contamination in waterways. One study has shown that 77 of 127 priority pollutants can be found in urban runoff (U.S. EPA, 1983). Long-term exposure to some of these substances can cause health problems in humans and other species that come into contact with the water. Short-term effects from waterborne pathogens are also a risk.

1.3 Flood Probability

The following are the most applied strategies used to outline flood possibilities:

- Statistical analysis of stream-flow records
- Regional methods
- Transfer methods
- Empirical equations
- Watershed modeling

In some municipalities, a few groups or rules require use of a specific technique. Many organizations and corporations have their own tactics. The goal of hydrologic observation for stormwater control is to outline the probability of flooding and the destiny of the flood events.

An effective plan to prevent flooding in an area includes

- highest rate of water flow—determines areas subject to flooding and helps design water transportation,
- runoff volume—helps design water storage facilitates, if needed, and
- time distribution of flow—helps design emergency plans or actions.

With small amounts of sample data and demand for a solution, it is hard to develop the most effective plan in predicting large flooding events. These predictions do not involve exact science; indeed, some data may conflict with information given about the area. Municipalities use previous flood events as examples to calculate the possibility of larger flood events to make decisions on whether or not to implement or improve the stormwater management of a specific area. The importance of this design factor shows how much to increase plausible events, not unfathomable ones. This helps with knowing roughly how much and where to allocate resources. This design factor also demonstrates why progress in water quality modeling cannot be measured without quantifying pollution. An accurate estimate of the flood harm potential is a key detail to a powerful flood damage abatement program.

2.0 INITIAL POLLUTION QUANTIFICATION AND PREDICTION

Initial pollution quantification and prediction demonstrates a need for stormwater, watershed, and receiving water quality modeling. Without quantifying pollution, there would be no way to measure progress of water quality modeling. The building blocks of pollution quantification and prediction, data collection and analysis, and scientific collaboration will be discussed in depth to establish a connection between pollution and water quality modeling.

There have been two notable collaborations regarding water quality. The first notable collaboration was that of the U.S. Water Resources Council, a collaboration between the federal government and the scientific community. The U.S. Water Resources Council laid a framework for what could be done regarding the analyzation of flood-flow frequency. In 1967, a standard for flood-flow frequency was published, titled *A Uniform Technique for Determining Flood-Flow Frequencies*. This guide recommends using log-Pearson Type III distribution (England et al., 2019). The guide was updated in 1976 to include an extension that introduced methods for outliers and presented the currently acceptable methods of analyzing peak-flow frequency data and gauging stations with sufficient detail to promote uniform application. Bulletin 17B, which was published in 1981 and editorially corrected in 1982, revised some of the techniques in previous editions and offers a further explanation of other techniques. Bulletin 17B perseveres with efforts to develop a coherent set of strategies used for defining flood potentials. However, the two goals of accuracy and consistency have not yet been fully attained.

The second notable collaboration is the collaboration of laws and regulations with stormwater management. Table 2.1 shows the laws passed to properly regulate stormwater in the United States.

3.0 EARLY MODELS

This section presents an overview of early model limitations due to scale and complexity, a discussion of the relationship between water quantity and quality. The early models of water quality laid the foundation for today's modeling capabilities.

3.1 Overview of Early Model Limitations

The use of water quality modeling in its early stages was limited because of scale and complexity. Early models could only be used for small-scale, simple scenarios because water quality models, like any other form of technology, gained sophistication with time; early models were simple because of the lack of available data and computation power. Water quality modeling has a relatively long history in the United States. The origins of water quality modeling can be traced back to the work of Streeter and Phelps in the 1920s. The practice of water quality modeling became more modernized in the 1960s when computers became more readily available. Progression continued by way of development in the 1980s with the growing popularity of the desktop computer. With the proliferation of the desktop computer came the increased use of geographic information system (GIS) technology. As computing became more robust, the Environmental Systems Research Institute, Inc., founded in 1969, released ARC/INFO, the first commercial GIS product, in 1981. The availability of GIS played an integral role in the advancement of model complexity.

TABLE 2.1 Legal and regulatory milestones for the U.S. Stormwater Program (National Research Council, 2009).

Year	Legal and regulatory milestone
1886	**Rivers and Harbors Act.** A navigation-oriented statute that was used in the 1960s and 1970s to challenge unpermitted pollutant discharges from industry.
1948 1952 1955	**Federal Water Pollution Control Act.** Provided matching funds for wastewater treatment facilities, grants for state water pollution control programs, and limited federal authority to act against interstate pollution.
1965	**Water Quality Act.** Required states to adopt water quality standards for interstate waters subject to federal approval. It also required states to adopt state implementation plans; although, failure to do so would not result in a federally implemented plan. As a result, enforceable requirements against polluting industries, even in interstate waters, was limited.
1972	**Federal Water Pollution Control Act.** First rigorous national law prohibiting the discharge of pollutants into surface waters without a permit. • Goal is to restore and maintain health of U.S. waters, • Protection of aquatic life and human contact recreation by 1983, • Eliminate discharge of pollutants by 1985, and • Water resource recovery facility financing.
1972	**Clean Water Act Section 303(d)** • Contains a water-quality-based strategy for waters that remain polluted after the implementation of technology-based standards. • Requires states to identify waters that remain polluted, to determine the total maximum daily loads that would reverse the impairments, and then to allocate loads to sources. If states do not perform these actions, U.S. EPA must.
1972	**Clean Water Act Section 208** • Designated and funded the development of regional water quality management plans to assess regional water quality, propose stream standards, identity water quality problem areas and identify wastewater treatment plan long-term needs. These plans also include policy statement that provide a common consistent basis for decision-making.
1977 1981	**Clean Water Act Sections 301 and 402** • Control release of toxic pollutants to U.S. waters. • Technology treatment standards for conventional pollutants and priority toxic pollutants. • Recognition of technology limitations for some processes.
1977	*NRDC vs. Costle.* Required U.S. EPA to include stormwater discharges in the National Pollution Discharge Elimination System (NPDES) program.

(*continued*)

TABLE 2.1 Legal and regulatory milestones for the U.S. Stormwater Program (National Research Council, 2009). (*Continued*)

Year	Legal and regulatory milestone
1987	**Clean Water Act Amended Sections 301 and 402** • Control toxic pollutants discharged to U.S. waters • Manage urban stormwater pollution • Numerical criteria for all toxic pollutants • Integrated control strategies for impaired waters • Stormwater permit programs for urban areas and industry • Stronger enforcement penalties • Anti-backsliding provisions
1990	**U.S. EPA's Phase I stormwater permits rules are promulgated** • Application and permit requirements for large and medium municipalities • Application and permit requirements for light and heavy industrial facilities based on Standard Industrial Classification (SIC) Codes, and construction activity ≥ 5 acres
1999	**U.S. EPA'S Phase II stormwater permit rules are promulgated** • Permit requirements for census-defined urbanized areas • Permit requirements for construction sites 1 to 5 acres
1997–2001	**Total Maximum Daily Load (TDML) program litigation** • Courts order U.S. EPA to establish TDMLs in a number of states if the states fail to do so. The TDMLs assign Waste Load Allocations for stormwater discharges that must be incorporated as effluent limitations in stormwater permits.
2006–2008	**Section 323 of the Energy Policy Act of 2005** • U.S. EPA promulgates rule (2006) to exempt stormwater discharges from oil and gas exploration, production, processing, treatment operations, or transmission facilities from NPDES stormwater program. • In 2008, courts order U.S. EPA to reverse the rule that exempted certain activities in the oil and gas exploration industry from stormwater regulations. In *Natural Resources Defense Council vs. EPA* (9th Cir. 2008) the court held that it was "arbitrary and capricious" to exempt from the Clean Water Act stormwater discharges containing sediment contamination that contribute to a violation of water quality standards.
2007	**Energy Independence and Security Act of 2007** • Requires all federal development and redevelopment projects with a footprint above 460 m^2 (5000 sq ft) to achieve predevelopment hydrology to the "maximum extent technically feasible".

Further advances occurred in the 1990s with the rise of the Internet. Without the Internet, the use/development of water quality models on a more complex level would have been nearly impossible, simply because the resources were not on-hand until technological advances paved the way. Early models did not have the benefit of the availability of GIS software and online databases, which are now used to promote much more realistic model parameters and application.

3.2 Water Quantity to Quality: "Dilution is the Solution to Pollution"

Water quality modeling was used to derive the common phrase, "dilution is the solution to pollution." Water quality modeling demonstrates the changes that would be evident if surface water runoff, carrying all of the contaminants acquired through traveling across urban settings, were to be diluted before large quantities of water were to make it to natural water sources. According to the Soil Science Society of America (n.d.), the way water moves in cities has large effects on both water quantity and water quality. In terms of water quantity, urbanization has the potential to cause two negative effects: (a) the presence of too much water and (b) not enough water reaching streams. Increased water quantities can cause greater economic costs, increased sedimentation, stress to wildlife, and shape changes to streams. In terms of water quality, runoff from urban settings includes the transference of contaminants such as metals, dirt, and debris. These contaminants either dissolve in the water or attach to the water molecules, with the exception of debris.

The basic categories of contaminants are commonly known as plant nutrients (nitrogen and phosphorus), metals (copper, zinc, and lead), organic chemicals related to gasoline, pathogens from animal feces, and trash/debris (commonly plastic). The level of contamination can be measured by the amount of total suspended solids. Poor water quality can lead to issues for both humans and wildlife, including beach closures, changes in behavior, and buildups of trash and debris. Using soil to filter surface water is a method that can potentially help the issue of pollution. Soil can be used to slow down runoff and dilute the water by removing suspended solids and dissolved contaminants before they reach natural water sources.

4.0 HISTORY OF POLLUTION REGULATIONS

The regulatory history of pollution plays a major role in the history of stormwater, watershed, and receiving water quality modeling. The use of water models in regulations in the United States is imperative to the history of pollution. Before pollution was regulated, there was no basis for water quality modeling. The increasingly polluted streams demonstrated the need to use models to mitigate the pollution. The increasing pollution also started a domino effect, sparking the attention of the public. With the rise of interest among the public, the government reacted by implementing regulations.

In the postwar period, the country's polluted water gained more attention. It was reported in 1945 that upwards of 3500 communities pumped more than 2 billion tons of raw wastewater into lakes, streams, and coastal waters every day. After this report, the Surgeon General warned that the population of the United States relied on water supplies of doubtful purity. The Federal Water Pollution Control Act of 1948 was accompanied by a report in which the Senate Committee on Public Works declared that "pollution of our water resources by domestic and industrial wastes has become an increasingly serious problem due to the rapid growth of our cities and industries.... Polluted waters menace the public health (through the contamination of water and food supplies), destroy fish and game life, and rob us of other benefits of our natural resources" (House Report no. 1829, to accompany Senate Bill 418, 80th Congress, 2d session, April 28, 1948). The Federal Water Pollution Control Act of 1948 was the first major U.S. law to address water pollution.

Further development of water quality modeling was driven in part by regulatory needs. The primary driver for water quality modeling in the United States was the Federal Water Pollution Control Act of 1956. There are four U.S. laws that address the environmental risks caused by toxic substances and their effect on watersheds. These laws regulate industrial chemicals, contaminated sites, hazardous waste, and pesticides. By the late 1970s, most conventional water quality processes and variables had been coded, followed by toxic variables and processes in the mid-1980s.

4.1 Clean Water Act of 1972

The Clean water Act (CWA) of 1972 was the result of growing public awareness and concern for controlling water pollution. The CWA came about from the major amendments to the Federal Water Pollution Control Act of 1948. According to U.S. EPA, the CWA establishes the basic structure for regulating discharges of pollutants into the waters of the United States and regulating quality standards for surface waters. The Federal Water Pollution Control Act of 1948 became widely known as the Clean Water Act of 1972 after the amendments that were made to the original act. There were notable amendments made in 1972, 1977, 1981, and 1987. Under the CWA, U.S. EPA was able to implement pollution control programs, including the setting of wastewater standards for industries. U.S. EPA was also able to develop national water quality criteria recommendations for pollutants in surface waters.

The implementation of the CWA made it illegal for any pollutant to be discharged from a point source in navigable waters without a permit being obtained. A "point source" can be defined as discrete conveyances such as pipes or human-made ditches, according to the U.S. EPA. The National Pollutant Discharge Elimination System (NPDES) program controls discharges. Permits from the NPDES program are required to be obtained by industrial, municipal, and other facilities that are not individual homes if their discharges go directly to surface waters.

Other notable laws that played a major role in the regulatory development of water quality modeling include the Toxic Substances Control Act; the Comprehensive Environmental Response, Compensation, and Liability Act; Resource Conservation and Recovery Act; and the Federal Insecticide, Fungicide, and Rodenticide Act. These laws address/regulate industrial chemicals, contaminated sites, hazardous waste, and pesticides, respectively.

4.2 Case Study: Chesapeake Bay Total Maximum Daily Load

The Chesapeake watershed extends over the District of Columbia and six mid-Atlantic States: Maryland, Virginia, Pennsylvania, New York, West Virginia, and Delaware. On December 29, 2010, U.S. EPA published the final Chesapeake Bay Total Maximum Daily Load (TMDL), which was to restore the bay water quality and aquatic resources to allow the region an invaluable recreational and economic resource. The Chesapeake Bay TMDL added wastewater treatment requirements to facilities throughout the watershed to clean up the identified pollutions of nitrogen, phosphorus, and sediment. Each watershed location was set with targets for the reduction of certain pollutants.

Excessive amounts of nitrogen and phosphorus were the leading factors in the impaired health of the bay. Figure 2.1 shows the leading sources of phosphorus and nitrogen into the Chesapeake Bay. Agriculture in both cases was the most important contributor, producing 42% of nitrogen flowing into the bay and 54% of phosphorus flowing into the bay. Other large contributors were urban stormwater and forests (University of Maryland, 2012).

In 2009, a comparison was done of the nutrition flow into the Chesapeake Bay among the six states in the bay watershed. Table 2.2 shows that 44% of the nitrogen into the bay originated in Pennsylvania, mainly coming from the Susquehanna River, which enters the bay at the northernmost point of Maryland. Of the large contributors of nitrogen flow, following Pennsylvania is Virginia with 27% and Maryland with 20%. The three states account for more than 90% of the nitrogen flow into the bay. The largest contributor of phosphorus was Virginia with 44%, followed by

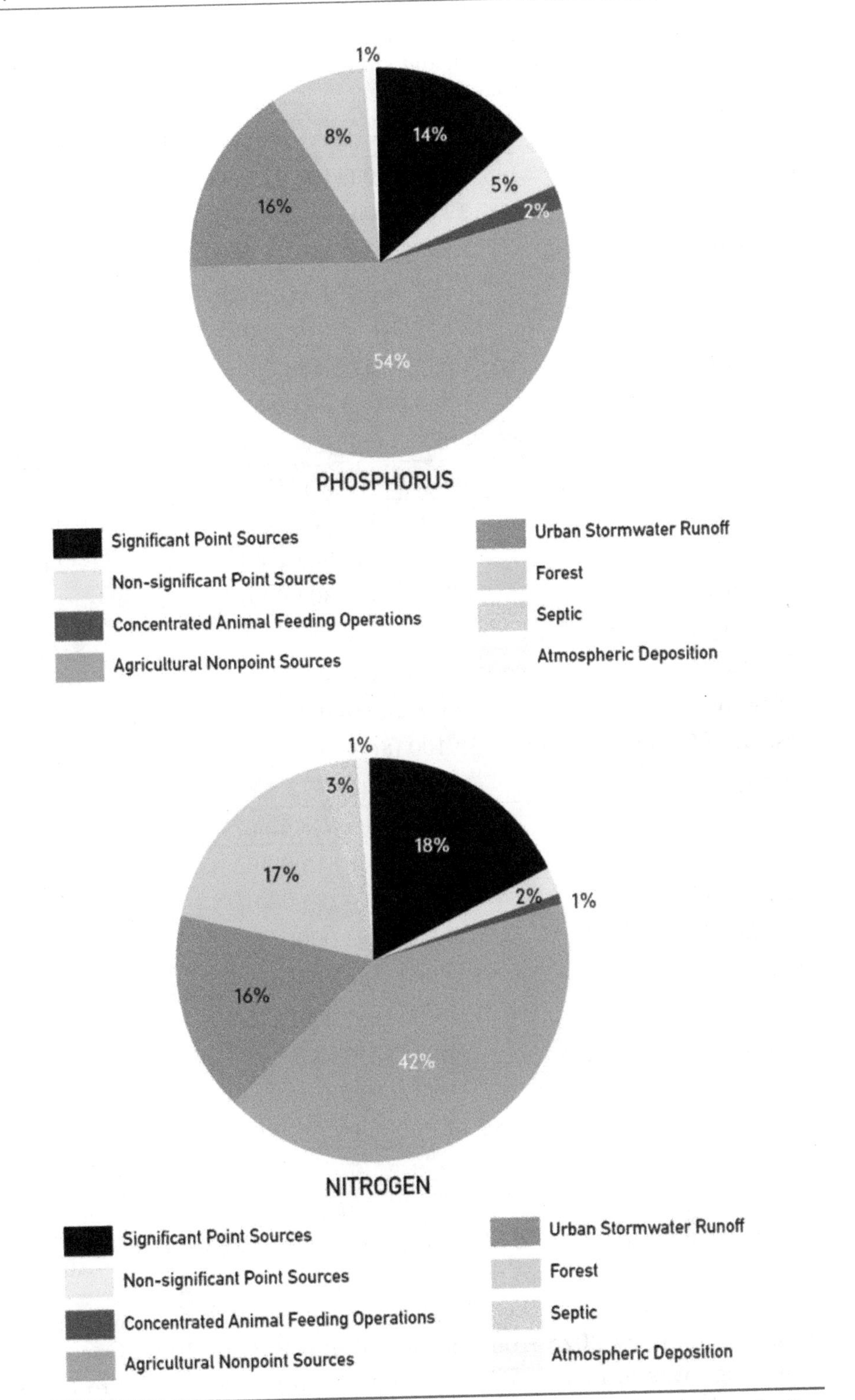

FIGURE 2.1 Leading sources of phosphorus and nitrogen in the Chesapeake Bay (University of Maryland, 2012).

TABLE 2.2 Nitrogen, phosphorus, and sediment flow into Chesapeake Bay from six states (University of Maryland, 2012).

State	Nitrogen (kg/yr [lb/yr])	Percent
Pennsylvania	48 268 100 (106 413 000)	44
Virginia	29 620 900 (65 303 000)	27
Maryland	22 417 000 (49 421 000)	20
New York	4 781 300 (10 541 000)	4
West Virginia	2 619 000 (5 774 000)	2
Delaware	1 896 000 (4 180 000)	2
District of Columbia	1 294 000 (2 853 000)	1
Totals	110 897 000 (244 485 000)	
	Phosphorus (kg/yr [lb/yr])	
Virginia	3 251 000 (7 168 000)	44
Pennsylvania	1 798 000 (3 965 000)	24
Maryland	1 499 000 (3 304 000)	20
West Virginia	—	5
New York	363 000 (801 000)	5
Delaware	143 000 (316 000)	2
District of Columbia	39 100 (86 400)	1
Totals	7 472 200 (16 473 400)	
State	**Sediment (tonnes/yr [tons/yr])**	**Percent**
Virginia	1 455 000 (1 616 000)	40
Pennsylvania	1 164 000 (1 283 000)	32
Maryland	629 000 (693 000)	17
West Virginia	170 000 (188 000)	5
New York	149 000 (164 000)	4
Delaware	29 300 (32 300)	1
District of Columbia	14 400 (15 900)	0
Totals	3 621 600 (3 992 200)	

Pennsylvania with 24% and Maryland with 20%. This accounted for 88% of the phosphorus flow into the bay.

The Chesapeake Bay TDML plan was set to seek a 38% reduction of all pollutants by 2025 from the agricultural level of 2009. Overall, the total design was to make 65% of the total nitrogen reductions to the bay over the time period. Behind nitrogen was wastewater, then stormwater. With Pennsylvania providing the largest percentage of nitrogen, most of the reduction will come from this state. Since 1985, Pennsylvania has been able to reduce nitrogen pollution by 27%; with the Chesapeake Bay TDML in action, they had composed a plan to continue their reduction. In 2011, Pennsylvania was able to have 40.43% of nitrogen pollution removed from the bay. The use of water quality modeling made it possible to conceptualize the percentages that needed to be changed to have an effect on the quality of water in the Chesapeake Bay area.

The Chesapeake Bay TMDL has been considered a "special case" when it comes to watersheds. According to U.S. EPA's website on the Chesapeake Bay TMDL, "EPA has full discretion to determine whether federal actions are appropriate based on the degree to which reduction goals are missed, the reasons why, and additional actions that jurisdictions are taking to ensure that load reductions will remain on track to meet the Partnership's goal of all practices in place by 2025 to meet applicable water quality standards. EPA has already demonstrated this discretionary authority when deciding whether to establish backstop allocations and adjustments in the Chesapeake Bay TMDL. The TMDL was to be implemented using an accountability framework that includes state and local watershed implementation plans (WIPs), two-year milestones, EPA's tracking and assessment of restoration progress and, as necessary, specific federal contingency actions, listed above, if the jurisdictions do not meet their commitments" (U.S. EPA, n.d.). In the *Chesapeake Bay TDML Executive Summary* (2010), U.S. EPA further states that, "[t]his accountability framework is being established in part to provide demonstration of the reasonable assurance provisions of the Chesapeake Bay TMDL pursuant to both the Clean Water Act (CWA) and the Chesapeake Bay Executive Order..." issued by former President Barack Obama in 2009 (U.S. EPA, 2010).

The Chesapeake Bay TMDL demonstrates a need for water quality modeling. Water quality modeling shines light on not only how much nitrogen and phosphorus are being dispersed into the Chesapeake Bay, but also from where these contaminants originate. Knowing the exact quantity of pollutants in the water will help to visualize the effects of the pollutants being released. This breakdown of pollutants will pinpoint how much of these contaminants are being contributed by each state to the Chesapeake Bay.

5.0 REFERENCES

Department of Energy and Environment. (n.d.). Why is stormwater a problem? Retrieved from https://doee.dc.gov/service/why-stormwater-problem

England, J. F., Bohn, T. A., Faber, B. A., Stedinger, J. R., Thomas, W. O., Veilleux, A. G. ... & Mason, R. R. (2019). A uniform technique for determining flood-flow frequencies (Bulletin 17C). *U.S. Geological Survey Techniques and Methods, 4* (B5), 148. https://doi.org/10.3133/tm4B5

National Research Council. (2009). *Urban stormwater management in the United States*. Washington, D.C.: The National Academies Press. https://doi.org/10.17226/12465.

School of Public Policy, University of Maryland. (2012). Saving the Chesapeake Bay TMDL: The critical role of nutrient offsets. Retrieved from http://faculty.publicpolicy.umd.edu/sites/default/files/nelson/files/660_--_environmental_workshop_report_final_spring_2012.pdf

Soil Science Society of America. (n.d.). Water quantity and quality. Retrieved from https://www.soils.org/discover-soils/soils-in-the-city/green-infrastructure/water-quantity-quality

U.S. Environmental Protection Agency. (n.d.). History of the Clean Water Act. Retrieved from https://www.epa.gov/laws-regulations/history-clean-water-act

U.S. Environmental Protection Agency. (2000). *The History of drinking water treatment* (EPA-816-F-00-006). Retrieved from U.S. EPA https://nepis.epa.gov/Exe/ZyPURL.cgi?Dockey=200024H9.TXT

U.S. Environmental Protection Agency [U.S. EPA]. (2010). Chesapeake Bay TMDL Executive Summary. Retrieved from https://www.epa.gov/sites/production/files/2014-12/documents/bay_tmdl_executive_summary_final_12.29.10_final_1.pdf

U.S. Environmental Protection Agency. (2012). *5.2 Dissolved oxygen and biochemical oxygen demand*. Retrieved April 2018, from https://archive.epa.gov/water/archive/web/html/vms52.html

6.0 SUGGESTED READING

Sebenik, P. G. (1975). Relationships of dissolved oxygen and biochemical oxygen demand in sewage effluent releases. The University of Arizona. Retrieved from https://repository.arizona.edu/handle/10150/554949

U.S. Environmental Protection Agency [U.S. EPA]. (2009). Source Water Protection Practices Bulletin Managing Stormwater Runoff to Prevent Contamination of Drinking Water. Retrieved from https://nepis.epa.gov/Exe/ZyPURL.cgi?Dockey=P1005P0Z.TXT

3

Evolution of Stormwater, Watershed, and Receiving Water Quality Modeling

Jagjit Kaur, Ph.D., Zachary Eichenwald, P.E., and Rajat Chakraborti, Ph.D.

1.0 EVOLUTION OF MODELS

The evolution of environmental modeling, including stormwater and water quality modeling, started from pre-computer era techniques to mainframes to user-friendly sophisticated tools that can run on laptops. A variety of physically based, empirical, and statistical methods/models have been developed by a number of different organizations and individuals working in science and academia. In the United States, the U.S. Environmental Protection Agency (U.S. EPA) has been the leader in the development and use of water quality models, whereas other federal and state agencies, such as the U.S. Army Corps of Engineers' Hydrologic Engineering Center and Waterways Experiment Station, have made strides in hydrologic and water quality modeling. Stormwater and water quality modeling is complex given its purpose to mathematically predict natural processes that are complex by definition (U.S. EPA, 2009). These methods and models enabled diagnostic evaluations of water quality problems and identification of effective management strategies for prioritizing actions. This chapter presents a historical account of stormwater and water quality modeling, a view of the current state, and a sketch of a possible future of stormwater, watershed, and water quality modeling.

The representation of flow of water and its constituents on land surface and subsurface environments has been going on for over 160 years, starting with Darcy's law discovered in 1856 and the Saint Venant equations describing unsteady open channel flow developed in 1871. A steady stream of analytical advances in the description of flow of water has occurred in the succeeding decades (Maidment, 1993). The origin of water quality modeling lies in the work of Streeter and Phelps in the 1920s. The evolution of models received a further boost with advances in computer technology and within environmental sciences. As documented in a review paper (Ambrose, 2008), the development of water quality modeling in the United States evolved through the following six broad periods:

1. Initial model development with mainframe computers (1960s to mid-1970s);
2. Model refinement and generalization with minicomputers (mid-1970s to mid-1980s);
3. Model standardization and support with microcomputers (mid-1980s to mid-1990s);
4. Better model access and performance with faster desktop computers running Windows and local area networks linked to the Internet (mid-1990s to early 2000s);
5. Model integration and widespread use of the Internet (mid-2000s to early 2010s); and
6. High resolution of data, cloud data storage, and real-time data integration (mid-2010s to present).

As Figure 3.1 shows, the development and application of water quality models can be traced with a succession of pollution problems, starting with the problem analysis and following with the models used to address those issues:

- Easily degradable organic wastes—addressed through dissolved oxygen and biochemical oxygen demand (BOD) models.
- Point-source discharges, non-point-source discharges, and eutrophication—addressed through nutrient cycle, phytoplankton, and ecological models.
- The nitrate problem—addressed through nitrogen cycle models.
- Toxics—addressed through food chain and ecological models.

The goal of these models was to understand the water pollution problems and develop solutions to manage those challenges. This goal continues today for the purpose of managing water resources by using more

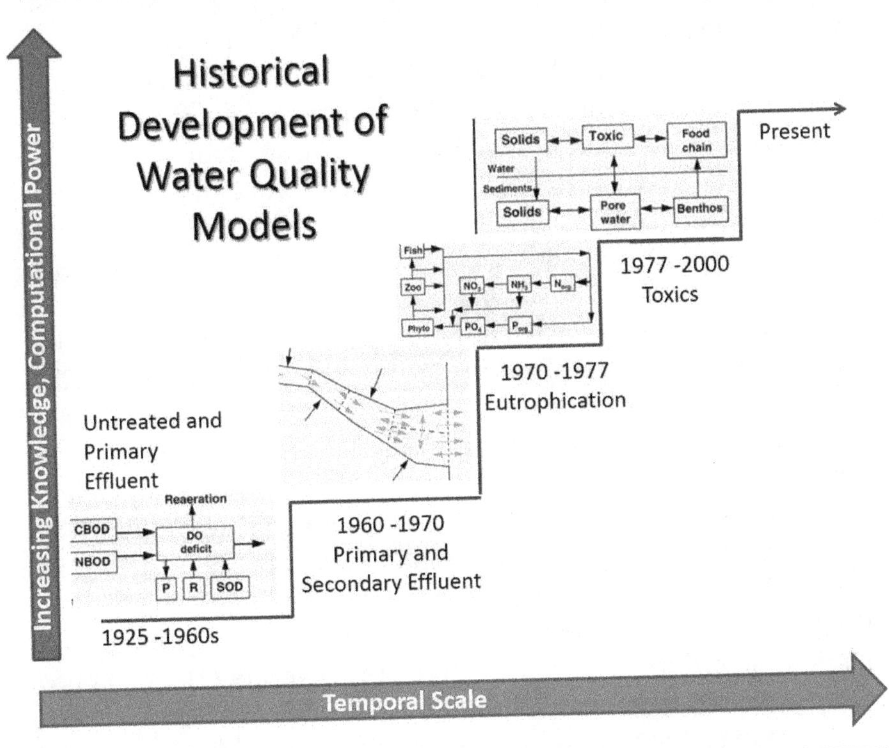

FIGURE 3.1 Timeline of evolution of water quality models (Chapra, 1997).

sophisticated tools and added technology with the application of models to determine the best tradeoff management solution where the level of water pollution is at the lowest possible level.

Although rainwater, or stormwater, has been considered as a resource in agricultural context for millennia, it was considered as a waste in urban context until recently (Echols & Pennypacker, 2015). The Clean Water Act (CWA) of 1972 was developed to protect rivers, streams, and lakes from pollution. As a part of the CWA, U.S. EPA created the National Pollutant Discharge Elimination System (NPDES) to track and control sources of pollution through permits. The U.S. EPA delegated the authority to issue and to enforce NPDES permits to the states.

Stormwater is also considered a source of pollution in urban areas because of the contamination it collects as it flows over impervious surfaces. Similar to the water quality models, the stormwater/watershed models represent the watershed hydrology and pollutant production including runoff, erosion, and washoff of pollutants from surfaces. The development of stormwater/watershed models also followed a similar pathway as the water quality modeling, starting from research tools and simple steady state models for localized areas to watershed/basin-scale tools that are used for water resources management (e.g., total maximum daily load [TMDL] development and stormwater management) with finer spatial and temporal resolutions. The goals of those models have been to simulate quantity and quality of runoff generated from land surfaces to develop strategies to prevent flooding, identify and stop illegal sources of pollution, and reduce stormwater runoff. With the advances in computer technology and computing power, integration of hydrologic/watershed models with receiving water quality models to run with high spatial and temporal resolutions has been made possible.

Statistical evaluations of precipitation volume, intensity, duration, and frequency have been performed over the past 100 years. The hydrologic modeling approach has evolved over time from single-event-based methods to continuous simulation methods. The continuous simulation approach is time consuming and costly and requires extensive datasets to develop, calibrate, and verify the models. The single-event models are simpler and involve simulating the rainfall-runoff process for a single storm event. As the understanding of the physical processes governing the rainfall-runoff models and computer programs for solving the mathematical equations describing the physical processes governing the generation of runoff by rainfall in watersheds were developed, a new era for finding solutions to manage stormwater problems on a watershed, or basin, scale has emerged.

2.0 DRIVERS OF MODEL EVOLUTION

The successive development of models has been driven by the availability of desktop computers, improved Windows operating systems, the Internet, and advances in science and technology. Model development has also been driven by the needs of regulations. With public awareness for the environment and impetus of regulations, linked watershed-water quality modeling has evolved to better assess the problems and develop management alternatives to find solutions to the problems. Drivers for model development include advances in technology and research, regulations, and other nonregulatory drivers to address planning, operational, and functional needs. The role of technology in the evolution of models can be specified in several ways, as described in the following subsections.

2.1 Computing and Communications

Before the introduction of relatively inexpensive personal computers, water quality models needed to be executed on a mainframe computer using punch cards. Computer time could be costly depending on the scope of a model, presenting a budgetary constraint on model application. For example, the Hydrologic Simulation Program FORTRAN (HSPF) Applications Guide lists the approximate cost to run a simple HSPF model on the Stanford cluster. The example model was made up of nine land segments and 13 stream reaches, which is a comparatively small model domain when compared with the typical scope of water quality models implemented today. A one-year simulation using this simple model implemented in HSPF 7.0 was estimated to cost $710 (1984 USD). Today, using a modern laptop computer and the Windows version of HSPF, this model would likely run in seconds at no cost to the user. Errors in model input would have been costly in the past; now, the user simply corrects the error and re-runs the model (U.S. EPA, 1984). The restrictions on computational power and the cost of running simulations was not limited to HSPF, and this significantly limited both the size and number of model simulations that could be run and the accessibility of these models to students and the public.

As computing power increased, and the overall cost of personal computers decreased, engineers and scientists began to replace the mainframe computer with desktop computers, and later laptop computers. The Windows operating system's graphical user interface (GUI) improved significantly along with this shift to personal computing-based modeling, allowing GUIs to be developed to interact with model code. The advent of these GUIs meant that the user did not need to interact with the model code solely through a text-based input file. For example, Figure 3.2 shows a comparison between U.S. EPA's Stormwater Management Model (SWMM) 4 (text-based) user interface and the SWMM 5 (GUI and text input) user interface. The SWMM 4 interface required the user to either be intimately familiar with the format of the text-based input file or to rely on the user manual to identify the nomenclature and location of each model parameter within the input file. In contrast to this approach, the modern SWMM user interface allows the user to input all model parameters directly into dialog boxes in a user-friendly environment.

Concurrent with the transition from mainframe computers to desktop computers, computing power and storage availability have been increasing. Although one-dimensional models are still widely used, the new computing power allows the application of 2- and 3-dimensional and dynamic frameworks. In addition, these significant improvements in processing

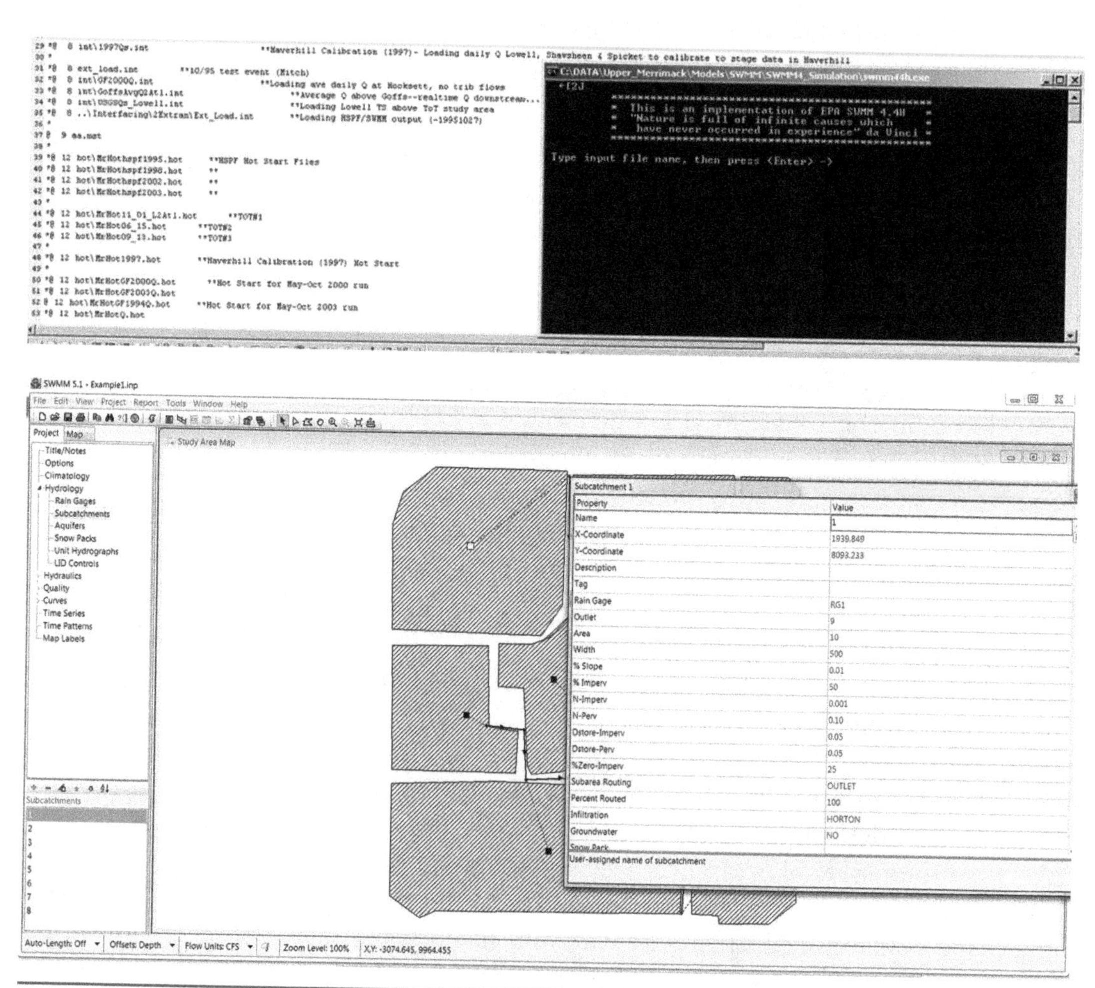

FIGURE 3.2 Comparison of the user interface for SWMM 4 (top) and SWMM 5 (bottom).

power have made more detailed modeling with various spatial and temporal resolutions possible. Now, hydrologic, hydraulic, and water quality models typically have hundreds, if not thousands, of unique elements. This increase (recall the relatively small size of the example the HSPF model run described above) allows greater detail and resolution in modeling and increases the scope and breadth of questions that can be answered using these models. Parallel processing can speed up model runs. For instance, U.S. EPA's SWMM has been re-coded to allow parallel processing, which can realize significant improvements in processing time for large, complex models. Along with the increased computing power, storage costs have been decreasing. Models typically now have output files in the gigabyte range (per model run).

2.2 Geographic Information System Software and Online Databases

The development of geographic information systems (GISs) and their use in stormwater, watershed, and receiving water quality modeling has proven to be a powerful tool to capture, store, manipulate, analyze, manage, and visualize geographical data spatially. This technology has provided a boost to modeling conducted by numerical models to organize data, conduct continuous time simulations with watershed configurations, and achieve high spatial and temporal resolutions. The development of model interfaces and user support tools occurred with the emergence of GIS technology in the 1960s.

The type of data used in models such as collection system network configuration, drainage configuration, quality of flow, hydrologic and water quality data, and watershed characteristics (e.g., land use, soils, slope, imperviousness), have improved over the years and provide more confidence in the predictive power of models. The following are descriptions of the data types and their statuses:

1. Both GIS and remote sensing data can be rapidly integrated for detailed representation of hydrologic and hydraulic systems, including drainage areas, subcatchment delineations, and watershed characteristics using detailed LiDAR or digital elevation models (Figure 3.3).
2. Flow metering and hydrologic data monitoring have significantly improved. Flow data can now be collected using relatively inexpensive, temporary metering technology that can be deployed anywhere from a sanitary or storm sewer to a large open channel. These data are routinely connected via a mobile Internet modem to transmit the data in near real time to the end user, which can both improve

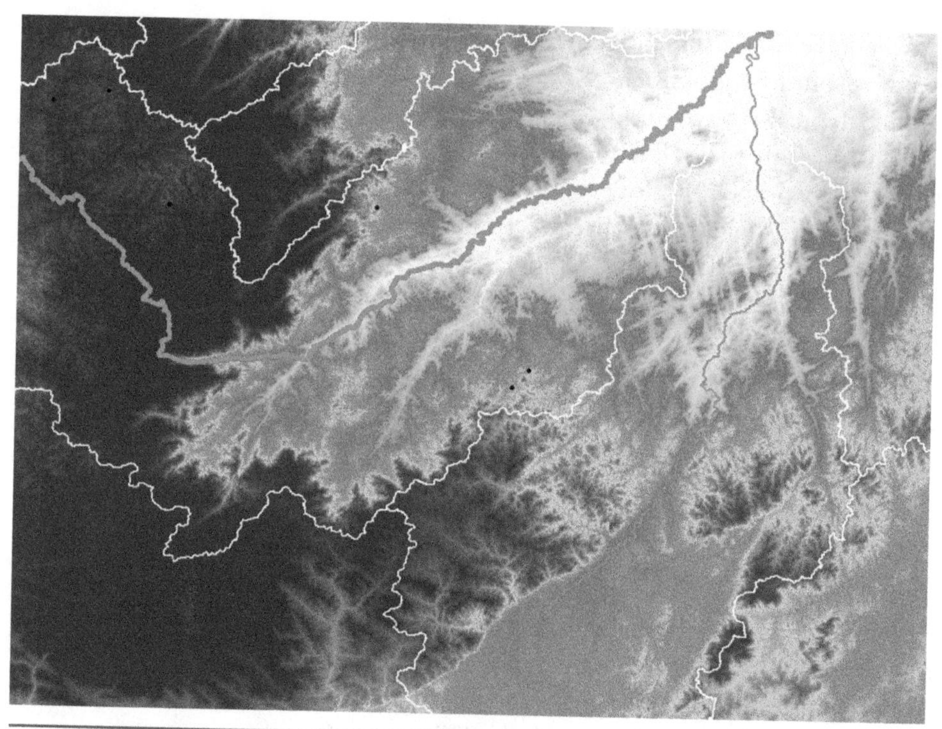

FIGURE 3.3 A GIS representation of hydrologic model subcatchments with LiDAR elevations superimposed.

efficiencies for remote sites and allow the data user to monitor system conditions in real time to identify adverse events that might require action. Data are often collected at fine resolution (e.g., 1 minute, 5 minutes, or 15 minutes).

3. Hydrologic, climatic, and water quality data are now readily available on the Internet, often free of charge to the end user. As an example, the United States Geological Survey (USGS) publishes all of its stream gauge data in real time on the Internet; any user with an Internet connection can view near real-time instantaneous data from any gauge throughout the country. These data can be used to calibrate and validate water quality models without needing to make a request to the data owner. If a user wishes to validate a model against a recent storm, all that is needed is a simple Internet query to the USGS database and the data are instantly available. Precipitation and other climatic data are available worldwide at daily, hourly, and sub-hourly intervals directly from the National Centers for Environmental Information, run by the National Oceanic and Atmospheric Administration (NOAA).

Water quality data are also readily available. For instance, the National Water Quality Monitoring Council offers surface water quality data from the USGS, U.S. EPA, and other state, federal, tribal, and local agencies. The National Stormwater Quality Database (http://www.bmpdatabase.org/nsqd.html) compiles and evaluates the water quality performance of best management practice (BMP) and low impact development (LID) installations (such as rain gardens and bioretention cells). These tools can be used to calibrate and validate water quality models and to represent water quality treatment, all potentially without collecting new monitoring data.

4. Enhancements in water quality monitoring technology change the type of data used for calibration. The following are a few examples of enhancements in monitoring technology:
   a. Auto-sampling technology can collect time-weighted or flow-weighted samples over the course of an event, giving a detailed time history of pollutant concentration and load for model calibration of the first flush and event mean concentration (Figure 3.4).

FIGURE 3.4 Autosampling device (USGS, 2005).

b. Enhanced in situ monitoring technology can measure continuous water quality parameters. Constituents that can be monitored range from those traditionally available (dissolved oxygen, temperature, conductivity) to new (chlorophyll a). Newer metering technology offers significant improvements in meter stability, requiring less frequent calibration and maintenance. Examples include optical dissolved oxygen (DO) metering technology, which replaces the membrane-based technology that was especially prone to meter drift and fouling.

c. Enhanced analytical chemistry methods can detect low levels of metals and organics, allowing more accurate quantification of the effects of pollutants in the environment. Analytical results are now routinely reported in units of parts per billion or parts per trillion. Examples of the importance of these increased analytical methods can be found in the 2009 U.S. EPA National Primary Drinking Water Regulations, which list the maximum contaminant level (MCL) for many regulated pollutants at the ppb or ppt level. For instance, endrin, a banned pesticide, has a MCL of 2 ppb, and ethylene dibromide, a petroleum byproduct, has an MCL of 50 ppt (U.S. EPA, 2009).

2.3 The Internet

Advances in Internet technology have furthered the development of stormwater, watershed, and receiving water quality models. These modeling web apps include availability of data, shared resources, and online tools/support systems, as follows:

- Many models are freely available for download from U.S. EPA, USGS, and other websites. Users can instantly access many widely applied hydrologic, hydraulic, and water quality models.
- Currently, GIS remote sensing data are widely available, including elevation data (DEM and LiDAR), imperviousness, land use, soils, and satellite imagery.
- The USGS has a freely available web-based application (StreamStats) to calculate drainage areas and other watershed characteristics at any point in the coverage area (Figure 3.5). Coverage currently extends to most U.S. states.
- Data can be transmitted directly to the user in near real time via telemetry. Data can include flow, stage, water quality, groundwater level, climate, and precipitation.

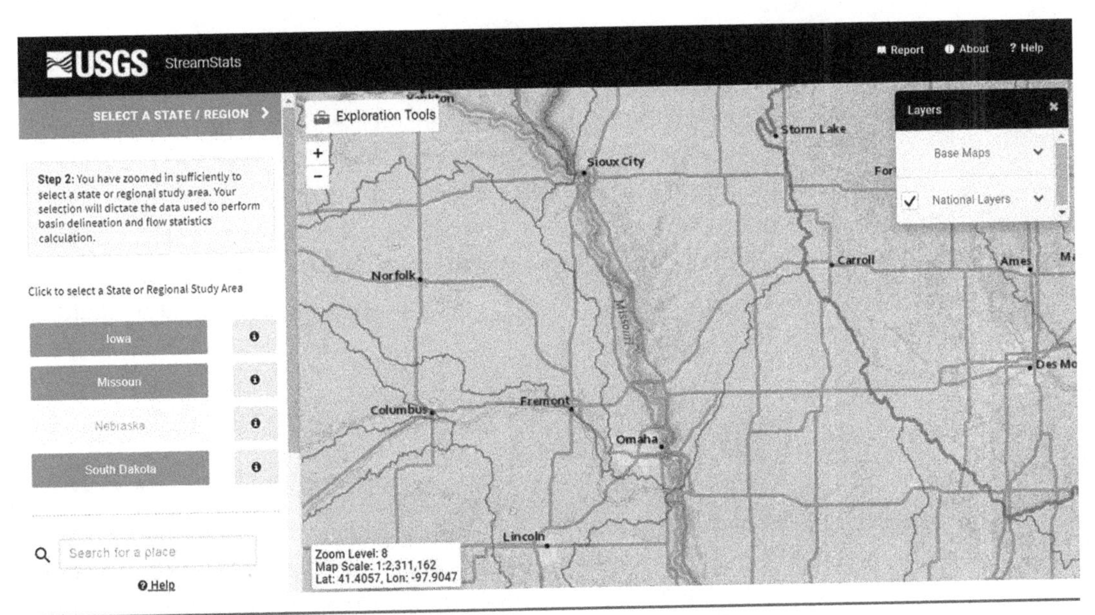

FIGURE 3.5 Screenshot of the USGS StreamStats web application.

- High-quality and quality-controlled climatic and hydrologic data are available from federal and state resources at no charge, dramatically simplifying model development, calibration, and validation efforts.
- Online tools for model development and application assistance allow modelers to connect with other expert users worldwide. Examples include U.S. EPA's SWMM-USERS listserv, which represents a broad community of engineers, scientists, and U.S. EPA model developers that collaborate to answer questions and suggest new features for the model code. In addition, users can find online references for many of the publicly available hydrologic, hydraulic, and water quality models that allow users to ask and receive answers to difficult modeling questions.
- Development of watershed-modeling web apps such as "Model My Watershed" represents a new way to provide an easy-to-use professional-grade modeling package to inform land-use decisions, support conservation practices, and enhance watershed education. The details of the model can be found at www.wikiwatershed.org.

2.4 Centralized Servers and Cloud Computing

Although personal desktop and laptop computers are extremely capable and are used for most model applications, some models are beginning to be implemented on remote computing resources (cloud-based model platforms).

These platforms can automatically download data from government sources such as USGS and NOAA, and they can link directly to a municipality's real-time supervisory control and data acquisition (SCADA) system to import data describing a collection system operation. This can allow real-time monitoring and modeling of combined sewer overflow (CSO) effects on receiving streams. Another key advantage of a cloud-based approach is that the data and any associated model output can be published online on a geospatially referenced user interface that can be offered to either key decision-makers or the public.

3.0 ROLE OF RESEARCH

Watershed hydrologic modeling research continues to modify and improve existing algorithms to represent the diverse physical, chemical, and biological processes (e.g., understanding the fate and transport of contaminants in waters and in food webs) to better define the environments and offer quantitative frameworks. These processes are affected by changed hydrology, erosion, sediment transport, surface water and groundwater interactions, watershed restoration approaches and methodologies such as use of green stormwater infrastructure, BMPs, and wetlands. An understanding of interactions among runoff generation processes, chemical and biological processes, and effect on various types of known and currently unknown pollutants (such as per- and polyfluoroalkyl substances [PFAS]) with runoff is crucial for water quality predictions using models.

Advancements in research in various areas such as understanding the fate and transport of various contaminants, modeling approaches, development of new methods for efficient and refined calculation of environmental processes to take advantage of increased processing power, modeling of BMPs in integrated modeling frameworks, and availability of new tools have assisted in the development of watershed, stormwater, and water quality models. More information on the latest research and development on the water environment can be found at the Water Research Foundation's website (https://www.waterrf.org).

4.0 ROLE OF REGULATIONS

Model development has also been driven by the needs of regulations. The primary regulatory driver for water quality modeling in the United States has been the Federal Water Pollution Control Act of 1956, including the CWA and amendments of 1972, 1977, 1981, and 1987. Four U.S. laws address environmental risks from toxic substances and their effect on watersheds. These laws are as follows:

- Toxic Substances Control Act (industrial chemicals),
- Superfund (contaminated sites),
- Resource Conservation and Recovery Act (RCRA) (hazardous waste), and
- Federal Insecticide, Fungicide, and Rodenticide Act (pesticides).

Model applications include the following:

- Models are used by federal and state regulators to set TMDLs. TMDLs may be used to set effluent limits. Examples include the Assabet and Blackstone rivers in Massachusetts, and many others nationwide.
- Models are used to determine available dilution and the effects of toxic pollutants on receiving water quality.
- Models are used to understand the fate and transport of contaminants in groundwater. This analysis can support cleanup projects funded under Superfund and RCRA.
- Models are used to establish the surface water cleanup goals.
- Models are used to determine the principal pollutant source (i.e., point). The model results can be used to determine how much pollutant reduction needs to attain water quality standards.
- The Federal Emergency Management Agency (FEMA) uses hydrologic and hydraulic models to predict the extent of the 100-year floodplain nationwide as a tool for flood control estimates. Models are used to evaluate the flood risk of properties and to determine whether flood insurance is necessary (FEMA, 2018). Estimation of flood risk relies on the models developed by the United States Army of Engineers' Hydrologic Engineering Center (HEC) model. Early models were designated HEC-1 (watershed hydrology) and HEC-2 (riverine hydraulics). Current versions of these models are HEC-HMS and HEC-RAS, which can now represent one-dimensional and two-dimensional steady state or time-varying hydraulics (USACE HEC, n.d.).
- U.S. EPA's SWMM was developed in the 1960s to aid in the planning and design of stormwater conveyance systems (Huber and Roesner, 2012). In this modeling framework, rainfall/runoff hydrology is coupled with detailed hydraulic routing. Continuous modeling is based on 1-dimensional solution of the Saint Venant equation. It allows modeling of both single-event (i.e., conveyance capacity) and long-term performance (i.e., storage capacity sizing).

5.0 OTHER DRIVERS

The water quality models, specifically stormwater, continue to evolve for the proper design of stormwater management approaches (such as BMPs) and hydraulic structures. Not only do the models help in understanding the problems and develop response relationships, but they also are used to evaluate the effectives of the treatment options.

Interaction between surface water (SW) and groundwater (GW) plays an important role in stormwater assessment. Integrated SW-GW modeling can provide a comprehensive and coherent understanding on basin-scale water cycle and better manage the water resources for sustainable usage. Of course, the GW-SW interaction should be treated differently at regional and local scales. Fully coupled (equations governing surface and subsurface flows are solved simultaneously within one software package) physics-based data-intensive models have great potential to meet the technical challenges (Barthel & Banzhaf, 2016). Loosely coupled schemes (i.e., two or more individual models are coupled via the exchange of model results, where the output of one model forms the input of the other) are also used for groundwater management projects depending on data availability and convenience to the "ready to use" models.

The models are most commonly used either as planning and decision aid for water management or to demonstrate compliance with regulations. The drivers of model development also include the following:

- Environmental protection awareness,
- Advances in water quality planning and pollution control,
- Advances in stormwater management using BMPs (such as LIDs) and stormwater green infrastructure facilities,
- Management goals, and
- Integrated frameworks/surface water groundwater interactions.

6.0 TRENDS IN WATER MANAGEMENT AND POTENTIAL DRIVERS FOR THE FUTURE

The evolution of models has reached an extremely important stage where technological advances have undergone a revolution over the last 50 years or so. In particular, application of GUIs, data storage capabilities, visualization of model outputs, and availability of high-resolution datasets under GIS systems have increased the model resolutions. Further, computer hardware and software advances have removed the computational constraints to develop models with fine spatial and temporal resolutions. Today, 2-dimensional and 3-dimensional models can be run in the order of minutes for conducting simulations with data of over 100 years. Significant advances in research have occurred to better characterize the natural systems with various physical, chemical, and biological processes to gain a deeper understanding of the ecosystems.

Since the 1920s, developers of the models have been trying to understand the functioning of natural ecosystems, fate and transport of pollutants, and cause/effect relationships. The goal had been to develop the historical trends, understand the present, and forecast the future trends. Today, and in the foreseeable future, more of the model applications are being generated to better manage the water systems and develop economical solutions to water quality problems.

Trends in water resources management are driven by stormwater management regulations, the nature of pollution problems, emerging contaminants, effects of extreme events, and development of holistic (integrated) approaches for sustainable management of water resources. Section 6.1 presents a summary of these evolving trends affecting the development of stormwater, watershed, and water quality modeling in the United States.

6.1 Stormwater Management Regulations

The stormwater regulations specified two types of stormwater permits: municipal and industrial. The CWA is the primary basis for all federal and state water quality programs. The Water Quality Act of 1987 provides the framework for the current regulations. The 1987 Stormwater Amendments created a structure encompassing different conceptual classifications of stormwater that might be contaminated and established different methodologies, including various types of discharge permits, to address each classification. On October 31, 1990, U.S. EPA issued the stormwater rule based on the 1987 Stormwater Amendments, which became effective on December 17, 1990. The classifications of stormwater systems subject to regulation included combined systems (i.e., sewer systems in which both sanitary wastewater and industrial process wastewater are mixed with rainwater and land runoff, primarily found in older urban areas); municipal separate stormwater systems (i.e., storm sewer systems owned or operated by municipalities that receive only stormwater runoff); separate stormwater systems (i.e., storm sewer systems that serve industrial facilities and were historically subject to or part of the industry's NPDES permit); and nonpoint-source runoff (i.e., all runoff that is not discharged to surface waters via a discrete pipe or conduit).

6.2 Changing Nature of Pollution Problems

Environmental changes including extreme events have caused many changes in the spectra of many pollution problems recently. Recent increased concentration of carbon dioxide and other "greenhouse gases" in the atmosphere and "acid rain" and its effects on lakes and streams, forests, and materials thinning of ozone in the stratosphere represent some of the changing nature of pollution problems. Studies have focused on the contamination and depletion of freshwater supplies on the depletion of the world's forests.

Other significant contributors are human activities that are implicated in detrimental environmental change and that are aimed at satisfying human needs and desires. These activities can only be expected to increase as the population grows. Human behavior can adversely affect the natural environment in a variety of ways and the forces that motivate environmentally detrimental behavior are likely to become even stronger in the future.

6.3 Emerging Contaminants and Pathogens

Up to 90% of oral drugs pass through the human body and end up in the water supply (Storteboom et al., 2010). Personal care products (such as soaps, cosmetics, etc.) also find their way into our water. Pharmaceuticals and personal care products (PPCPs) and endocrine disrupting compounds (EDCs) are among the prime examples of emerging contaminants that are increasingly being detected at low levels in surface water, and there is concern that these compounds may have an effect on aquatic life (Water Quality Association). These chemicals are known as "contaminants of emerging concern" or "emerging contaminants" because they had not previously been detected and are being discovered in trace amounts in the water supply. Emerging contaminants are important because the risk they pose to human health and the environment is not yet fully understood. The emerging contaminants may also demonstrate low acute toxicity, but cause significant reproductive effects at very low levels of exposure. In addition, the effects of exposure to aquatic organisms during the early stages of life may not be observed until adulthood (U.S. EPA). Endocrine disruptors are substances that may interfere with the function of hormones in the body. Trace amounts of these contaminants are being discovered in water throughout the country (Water Quality Association).

Another emerging contaminant of concern is PFAS. Water treatment systems are not fully developed to remove these compounds efficiently, and the analysis technique for these compounds needs advancements, particularly when they are present in trace amounts.

Emerging pathogens can be found in a wide range of natural and artificial environments. Legionella is an environmental pathogen that causes disease when aerosolized droplets containing the bacteria are inhaled. Legionella causes two diseases in humans: a severe pneumonia (Legionnaires' disease) and Pontiac fever (cold-like symptoms) (Jang et al., 2014). Emerging contaminants and pathogens are becoming drivers of stormwater projects to understand their potential effects on human health and the environment.

6.4 Optimization to Develop Economical Solutions

The solution of problems related to stormwater management require decision-making and selection between a number of alternatives that need to satisfy a number of technical and regulation constraints. A significant number of scientific works have appeared in the literature dealing with the environmental issues through the development of optimization models and their implementation in practical cases (Kondili, 2005). The type of optimization technique is variable with identification of the basic problem parameters of either deterministic and stochastic programming models. Progress in linear and integer programming solvers and software tools over the past decades have meant more reliable and rapid solution of even large-scale problems (Boland, 2009).

6.5 One Water Holistic Approach to Manage Water Systems

Effective management of water resources is the solution to many uncertainties. The concept of One Water shown in Figure 3.6 is the holistic (integrated) approach to manage water systems where the science of water advances all portions of the water cycle. Many benefits are realized when the barriers traditionally separating water, wastewater, stormwater, and water reuse are broken down in a reliable and resilient plan for life's most essential resource. The principle idea behind the One Water planning process includes (a) all water has value, (b) a focus on achieving multiple benefits, (c) watershed-based approach, (d) partnership for progress, and (e) inclusion and engagement of all.

In order to integrate wastewater facilities planning with stormwater, recycled water, and water conservation with a long-term planning horizon, coupled watershed and water quality modeling can be developed with emphasis on the effect on receiving waters. One Water planning is a holistic approach that has been gaining momentum to develop a more integrated approach for meeting TMDLs of stormwater discharges (City of Los Angeles, 2018).

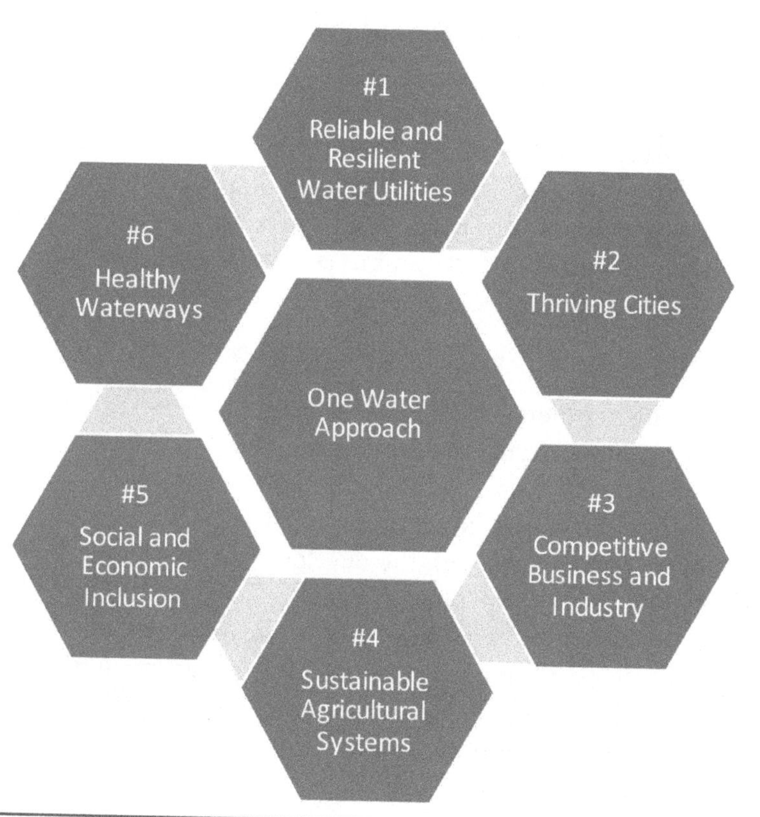

FIGURE 3.6 The framework of the One Water planning approach (redrawn from U.S. Water Alliance, 2017).

6.6 Extreme Events Affecting Stormwater Quality and Quantity

As summarized by Sharma et al. (2016), stormwater runoff from impervious areas such as roads, roofs, and parking lots is increasingly catching attention because of effects such as increased frequency of flooding, deterioration of water quality in receiving waters, and risk to downstream ecosystems (Kayhanian et al., 2008; McQueen et al., 2010; Milly et al., 2002; Walsh, Fletcher, & Burns, 2012). Stormwater runoff contains a range of pollutants in dissolved and particle-bound forms. The adverse effect of a substance depends on properties like persistence, toxicity, and bioaccumulation. These features are affected by the partitioning of the substance between the dissolved and particulate phases as well as by the exposure and uptake mechanism of the type of flora/fauna considered. For example, hydrophobic organic compounds and heavy metals can sorb to suspended solids carried with stormwater runoff, accumulate in sediments of receiving waters, and negatively affect benthic organisms (Nakajima et al., 2006), and compounds freely available in the water phase such as dissolved copper can lead to direct toxic effects to water living organisms (Allen & Hansen, 1996; Ma, Kim, Allen, & Cha, 2002).

Management of stormwater quality is an essential part of strategies to improve the environmental status of natural waters. Most design practices for constructing stormwater management facilities are, however, based on historical climate conditions (Sharma et al., 2016). Climate predictions show that the future conditions will be different from those of the past, with an expected increase in the frequency of heavy rainfall events and droughts in the future in some regions (IPCC, 2012; Parry et al., 2007). Increased intensity of heavy rain storms and longer dry weather periods are some of the important expected climate change effects. Increasing occurrences of precipitation-extreme events for short time scales are likely to occur for return periods of 0.5 years and higher in many parts of the world. These changes may affect stormwater runoff quality (He, Valeo, Chu, & Neumann, 2011; Wilson & Weng, 2011) and the efficiency of stormwater treatment systems because longer dry weather periods may lead to increased buildup of sediments on catchment surfaces and thus higher concentrations and load pulses in the runoff. This may be exacerbated by the more intense heavy rain storms.

The high rain-induced runoff after a long gap of rainfall often causes high pollutant load to the stormwater because of "first flush" effects. High flows may further disrupt the settling process and shorten the hydraulic residence time (HRT) of stormwater retention ponds during extreme conditions, which may lead to higher pollution concentrations and loads being discharged to the environment. These effects should, therefore, be taken into account during the selection of stormwater pollution control strategies as part of surface water protection plans so that climate-change-resilient solutions can be implemented (Charlesworth, 2010). This can be achieved by applying integrated stormwater quality models in combination with analysis of climate change scenarios. Hathaway, Brown, Fu and Hunt (2014) conducted such studies focusing on the hydrological behavior of stormwater control measures.

Some of the constituents affected by extreme events are total suspended solids and copper (total and dissolved) loads and concentrations in runoff from a catchment as well as in the outlet of a stormwater retention pond. Stormwater management model inputs include the rainfall time series and the pollutant fluxes released by sources in the catchment.

6.7 Changing Nature of Pollution Problems

Global environmental threats such as more frequent extreme events resulting in huge quantities of runoff drawing more and more pollutants into the stormwater system, threat of the problems caused by emerging contaminants and pathogens, and the mass extinction of species are essential to assessing the significance of these threats. For example, perfluorooctanoic acid (PFOA) and perfluorooctane sulfonic acid (PFOS) were first manufactured in the 1950s as effective fire suppression agents. Industry has moved away from certain PFAS compounds, namely PFOA and PFOS, over concerns that they are harmful to human health, replacing them with what are believed to be less toxic PFAS compounds. Availability of sophisticated and sometimes expensive analysis methods is helping to track and regulate those pollutants in the stormwater system. It has been a constant evolution of analysis methods, regulations, and implication of management methods due to pollutants' changing nature.

7.0 SUMMARY AND CONCLUSION

The stormwater, watershed, and receiving water quality field evolved over decades and many models have been developed (refer to Chapter 4 for model details). This evolution occurred in many facets, ranging from steady state models to dynamic models, from point-source models to coupled models from point- and non-point sources, 1-dimensional models to 3-dimensional models, single systems to integrated frameworks, and so on to understand the effect on land surface on downstream water quality in receiving waters, technology advancements affecting data gathering, manipulation, data storage, visualization, processing speed, and ease of use. These models are powerful tools to gain valuable knowledge to predict changes in the water quality of a system based on physical, chemical, and biological changes within watersheds and in waterbodies. However, challenges remain to make proper and fruitful use of the models using fast-paced development in science and technology, as described in this chapter. The nexus of stormwater, watershed, and water quality modeling can provide a holistic view of hydrological-relating terrestrial processes to water quality benefits to provide ecosystem services (Figure 3.7).

The changes made on land affect the hydrological cycle and stormwater quality generated from land surface affecting aquatic ecosystem and ecosystem services. In particular, consideration of data and use of data in various temporal and spatial scales could be given more attention to make the decision of selecting one particular model over another, including the uncertainty in their prediction capabilities. A big challenge still remains when it comes to using these models in various countries as standardization will be needed.

Recent advancements in the development of Internet-based frameworks that provide significant cost savings for the management and application of

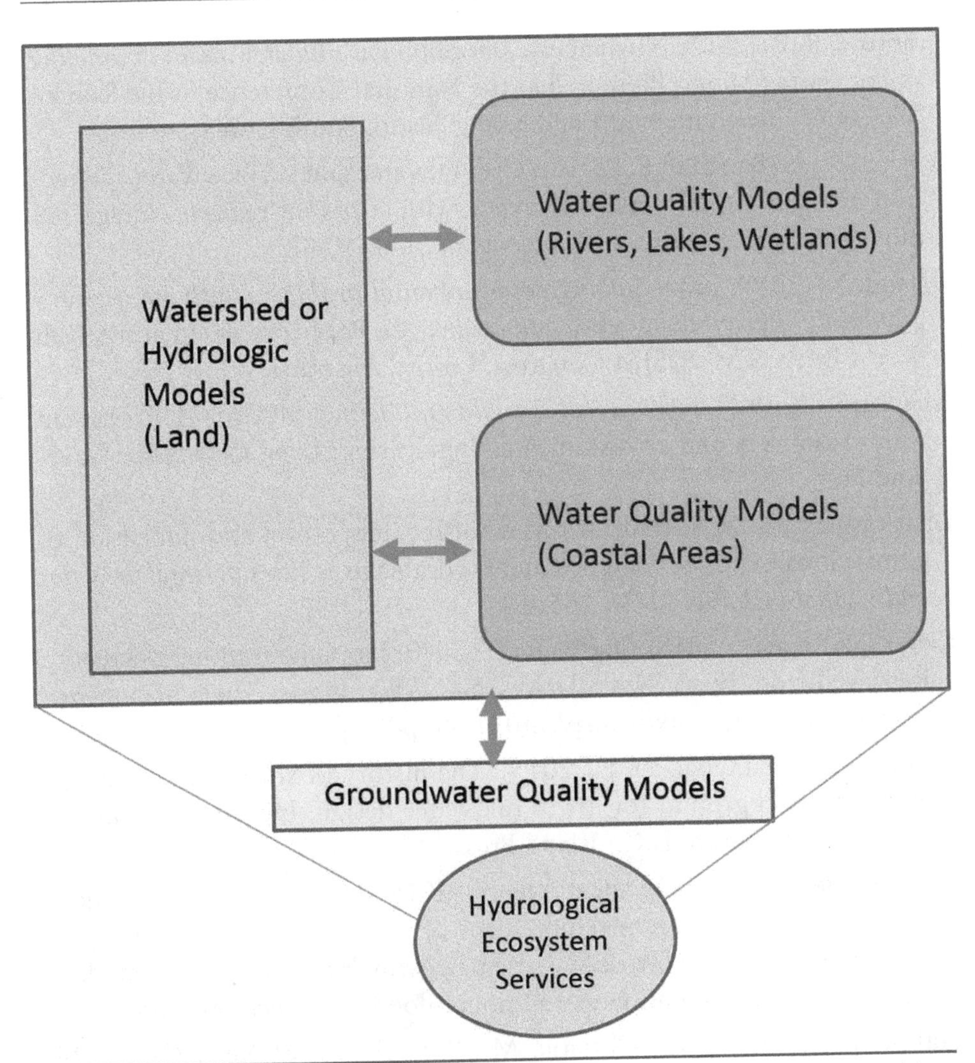

FIGURE 3.7 Nexus of stormwater, watershed, and receiving water quality modeling (redrawn from Hallouin et al., 2018).

models are a step forward in the next generation of watershed and water quality models. However, the research requires continuous testing and updates to the models, and developing linkages and standardized tools to best mitigate watershed and water quality issues remains paramount to these developments.

8.0 REFERENCES

Allen H. E. & Hansen D. J. (1996). The importance of trace metal speciation to water quality criteria. *Water Environment Research*, *68*(1), 42–54.

Ambrose, R. B. (2008, November). *Development of water quality modeling in the United States*. Presented at the National Conference of the Korean Society of Environmental Engineering, Seoul, South Korea.

Barthel, R., & Banzhaf, S. (2016). Groundwater and surface water interaction at the regional-scale—A review with focus on regional integrated models. *Water Resources Management, 30*(1), 1–32.

Boland, N. (2009, July). *Solving environmental problems with integer programming: Recent experience and challenges*. Paper presented at the 18th World IMACS/MODSIM Congress, Cairns, Australia.

Chapra, S. C., 1997. *Surface water-quality modeling*. McGraw-Hill series in water resources and environmental engineering. Long Grove, IL: Waveland Press.

Charlesworth, S. M. (2010). A review of the adaptation and mitigation of global climate change using sustainable drainage in cities. *Journal of Water and Climate Change, 1*(3), 165–180.

City of Los Angeles. (2018). One Water LA 2040 plan: Guiding principles report. Retrieved from https://www.lacitysan.org/cs/groups/sg_owla/documents/document/y250/mdmx/~edisp/cnt031540.pdf

Echols, S., & Pennypacker, E. (2015). The history of stormwater management and background for artful rainwater design. In, *Artful Rainwater Design*. Washington, D.C.: Island Press.

Federal Emergency Management Agency. (2018). Hydraulic numerical models meeting the minimum requirement of national flood insurance program. Retrieved from https://www.fema.gov/hydraulic-numerical-models-meeting-minimum-requirement-national-flood-insurance-program.

Hallouin, T., Bruen, M., Christie, M., Bullock, C., & Kelly-Quinn, M. (2018). Challenges in using hydrology and water quality models for assessing freshwater ecosystem services: A review. *Geosciences, 8*(2), 45. DOI:10.3390/geosciences8020045

Hathaway, J. M., Brown, R. A., Fu, J. S.; & Hunt, W. F. (2014). Bioretention function under climate change scenarios in North Carolina, USA. *Journal of Hydrology, 519*(A), 503–511.

Huber, W., & Roesner, H. (2012). *The history and evolution of the EPA SWMM*. Presented at Fifty Years of Watershed Modeling: Past, Present and Future, Boulder, CO.

Intergovernmental Panel on Climate Change. (2012). Managing the risks of extreme events and disasters to advance climate change adaptation. In Field, C. B., Barros V., Stocker, T. F., Qin, D., Dokken, D. J., Ebi, K. L., Mastrandrea, M. D., Mach, K. J., Plattner, G.-K., Allen, S. K., Tignor, M.,

Midgley, P. M., (Eds.), *A Special Report of Working Groups I and II of the Intergovernmental Panel on Climate Change*. Cambridge, U.K.: Cambridge University Press.

Jang, G., Smith, M., Linton, K., & Blaha, F. (2014). Emerging pathogen of concern. Edited by Jen Clancy (Ed.), *Advances in Water Research*. Retrieved from http://www.waterrf.org/resources/magazine/documents/awrvol24no4.pdf

Kayhanian, M., Stransky, C., Bay, S., Lau, S.-L., & Stenstrom, M. K. (2008). Toxicity of urban highway runoff with respect to storm duration. *Science of the Total Environment, 389*(2–3), 386–406.

Kondili, E. (2005). *Review of optimization models in the pollution prevention and control*. Paper presented at the European Symposium on Computer Aided Process Engineering—15; Puigjaner, L., & Espuña, A., (Eds.), Amsterdam, Netherlands: Elsevier Science B.V.

Ma, H., Kim, S. D., Allen, H. E., & Cha, D. K. (2002). Effect of copper binding by suspended particulate matter on toxicity. *Environmental Toxicology & Chemistry, 21*(4), 710–714.

Maidment, D. R. (1993). *Handbook of Hydrology*, New York, NY: McGraw-Hill.

McQueen, A. D., Johnson, B. M., Rodgers, J. H., Jr., & English, W. R. (2010). Campus parking lot stormwater runoff: Physicochemical analyses and toxicity tests using *Ceriodaphnia dubia* and *Pimephales promelas*. *Chemosphere, 79*(5), 561–569.

Milly, P. C. D., Wetherald, R. T., Dunne, K. A., & Delworth, T. L. (2002). Increasing risk of great floods in a changing climate. *Nature, 415*, 514–517.

Nakajima, F., Saito, K., Isozaki, Y., Furumai, H., Christensen, A. M., Baun, A.; Ledin, A., & Mikkelsen, P. S. (2006). Transfer of hydrophobic contaminants in urban runoff particles to benthic organisms estimated by an in vitro bioaccessibility test. *Water Science & Technology, 54*(67), 323–330.

Parry, M. L., Canziani, O. F., Palutikof, J. P., van der Linden, P. J., & Hanson, C. E. (2007). Contribution of working group II to the fourth assessment report of the Intergovernmental Panel on Climate Change 2007. (pp. 23–78) Cambridge, U.K.: Cambridge University Press.

Storteboom, H., Arabi, M., Davis, J. G., Crimi, B., & Pruden, A. (2010). Tracking antibiotic resistance genes in the South Platte River Basin using molecular signatures of urban, agricultural, and pristine sources. *Environmental Science & Technology, 44*(19): 7397 DOI: 10.1021/es101657s

U.S. Army Corps of Engineers. (n.d.). About HEC—History. Retrieved from https://www.hec.usace.army.mil/about/history.aspx

U.S. Environmental Protection Agency. Contaminants of emerging concern including pharmaceuticals and personal care products. Retrieved May 2, 2018, from https://www.epa.gov/wqc/contaminants-emerging-concern-including-pharmaceuticals-and-personal-care-products

U.S. Environmental Protection Agency. (1984). Application guide for Hydrological Simulation Program – FORTRAN (HSPF) (EPA-600/3-84-065). Washington, D.C.: U.S. Environmental Protection Agency.

U.S. Environmental Protection Agency. (2009). National primary drinking water regulations (EPA-816/F-09-004). Washington, D.C.: U.S. Environmental Protection Agency.

U.S. Geological Survey. (2005). Nutrient loading at two coastal creeks in Everglades National Park. Poster presented at the Florida Bay and Adjacent Marine Systems Science Conference, Duck Key, Florida. Retrieved from USGS archives https://archive.usgs.gov/archive/sites/sofia.usgs.gov/publications/posters/nutloading_coastalcreeks/

U.S. Water Alliance. (2017). *One Water for America policy framework: Executive summary*. One Water for America Listening Sessions. Retrieved from http://uswateralliance.org/sites/uswateralliance.org/files/publications/One%20Water%20for%20America%20Policy%20Framework%20Executive%20Summary.pdf

Walsh, C. J., Fletcher, T. D., & Burns, M. J. (2012). Urban stormwater runoff: A new class of environmental flow problem. *PLOS ONE*, 7(9), e45814.

Water Quality Association. Contaminants of emerging concern. Retrieved from https://www.wqa.org/whats-in-your-water/emerging-contaminants

Wilson, C. O., & Weng, Q. H. (2011). Simulating the impacts of future land use and climate changes on surface water quality in the Des Plaines River watershed, Chicago Metropolitan Statistical Area, Illinois. *Science of the Total Environment, 409*(20), 4387–4405.

4

Today's Models

Sabu Paul, Ph.D., P.E., PMP, Mahmudul Hasan, Dmitrijs Obolevics, IEng MICE, Gian Villarreal, P.E., CFM, CPSWQ, and Rishab Mahajan

1.0 INTRODUCTION

Today's water quality models are broadly categorized into two groups: watershed models and instream or receiving water models. Watershed models are the models that deal with watershed hydrology and quality. They address the hydrologic cycle at the watershed level and pollutant production at that stage, including runoff, erosion, and washoff of sediment and other pollutants. The instream or receiving water quality models emphasize hydrology and water quality of water conveyance systems, such as rivers, reservoirs, lakes, and estuaries. Only a handful of models are capable of handling the processes at watershed and receiving waterbodies' levels. However, systems are developed to link watershed models to receiving water models.

In this review, the models are grouped as watershed quality models, urban stormwater quality models, receiving water quality models, integrated modeling systems, and water quality compliance models. The stormwater quality models are those models that address stormwater quantity and quality in primarily an urban environment, whereas watershed quality models (in this review) are those that also handle rural or agricultural environments to a certain extent. The integrated modeling systems consist of a combination of individual models or tools that provide linkages between them to address a broader domain. The water quality compliance models are those models that are currently used for meeting regulatory stormwater compliance.

Although the models can be categorized into these different groups, most of the models will not fall exclusively into a single category. Some of the models can be considered as both watershed models and instream models. In fact, most of the watershed models include an internal river transport and water quality processes in reservoirs. However, the processes are simplified compared to the more complex receiving water models. Some of these can also be included in the urban stormwater quality models. For example, the Hydrologic Simulation Program FORTRAN (HSPF) model is one of the models that handles watershed processes very well with some capabilities of a receiving water model, along with the option to simulate stormwater quality and the ability to help with stormwater compliance (best management practice [BMP] simulations). However, BMPTrains is a model that can be used for stormwater compliance that also handles some of the watershed processes. Because many models fall into multiple categories, Table 4.1 provides the primary grouping under which the model descriptions are provided in this review.

The models can also be categorized based on their abilities to handle different pollutants, such as sediment, nitrogen, phosphorus, water temperature, bacteria, pesticides, metals, and dissolved oxygen. Table 4.1 also lists the pollutants that are simulated by the selected models.

2.0 WATERSHED QUALITY MODELS

This section describes models that address hydrology and water quality at the watershed level that can handle both the urban and rural environment at varying levels. Some of them are better suited for a rural environment but are included here because they are widely used public domain models. The watershed models can be further classified into various categories based on (a) their ability either to handle overland processes only, or also be able to map the flow of pollutants through rivers or other conveyance systems, (b) their complexity in estimating the watershed loads, and (c) the resolution of the prediction in terms of timestep (i.e., daily or sub-daily loads). Some models simulate only surface hydrology, whereas the complex models take into consideration the effect to groundwater. The models also differ in their ability to simulate different pollutants. Borah, Ahmadisharaf, Padmanabhan, Imen, and Mohamoud (2019) provide a detailed description of 14 watershed models, including their capabilities, credibility, and suitability in total maximum daily load (TMDL) development and implementation.

2.1 ANSWERS

ANSWERS is a commonly used watershed quality model comprising two major components: hydrology (Huggins & Monke, 1966) and upland erosion (Foster & Meyer, 1972). This model is designed for ungauged watersheds and for evaluating the effectiveness of agricultural and urban watershed BMPs (Parsons et al., 2004). In this model, hydrologic and water

TABLE 4.1 The categorization of models based on their capabilities.

Model	Watershed	Receiving Water	Stormwater	Stormwater Compliance	Integrated System	Primary Grouping[a]	Pollutants Addressed[b]
AGNPS	•	–	–	–	–	W	N, P, S, B, Pc
ANSWERS	•	–	–	–	–	W	S
BASINS	•	•	•	•	•	I	T, DO, N, P, S, B, Pc, M
BATHTUB	–	•	–	–		R	N, P, A
BMPTRAINS	○	–		•	–	C	N, P
CE-QUAL-ICM	–	•	–	–	–	R	N, P, M, A
CE-QUAL-RIV1	–	•	–	–	–	R	T, DO, N, P, B, M
CE-QUAL-W2	–	•	–	–	–	R	T, DO, N, P, S, B, Pc, M
CH3D-WES	–	•	–	–	–	R	T, DO, N, P, S, B
CORMIX	–	•	–	–	–	R	T, DO, S, B, M
Delft3D	–	•	–	–	–	R	T, DO, N, P, S, B
EFDC	–	•	–	–	–	R	T, DO, N, P, S, B, Pc, M
EUTROMOD	•	•	–	–	–	W	N, P, A
GWLF	•	–	–	–	–	W	N, P, S, B
HEC-RAS	–	•	–	–	–	R	T, DO, N, P, S, A
HSPF/LSPC	•	•	•	•	–	W	T, DO, N, P, S, B, Pc, M
IDEAL	•			•		C	N, P, S, B
National Stormwater Calculator	•	–	•			S	
P8-UCM	•	–	•			S	N, P, S
Visual PLUMES		•				R	B
QUAL2E	–	•	–	–		R	T, DO, N, P, S, B, Pc, M, A
QUASAR	–	•				R	DO, N
Regression Method	○					W	
RIVMOD-H		•				R	
Simple Method	○					W	
SLAMM/WinSLAMM	•	–	•	•		W	N, P, S, M
SUSTAIN	•	•	•	•	–	C	N, P, S, B, Pc, M
SWAT	•	•	•	•		W	T, DO, N, P, S, B, Pc, M

(continued)

TABLE 4.1 The categorization of models based on their capabilities. (*Continued*)

Model	Watershed	Receiving Water	Stormwater	Stormwater Compliance	Integrated System	Primary Grouping[a]	Pollutants Addressed[b]
SWMM	•	•	•	•		S	N, P, S, B, Pc, M
Tidal Prism Model		•				R	
TOXMOD		•				R	
U.S. EPA Screening Procedures	○		–	–	–	W	
VRRM	•			•		C	N, P
WARMF	•	–				W	T, DO, N, P, S, B, Pc, M
WASP	–	•	–	–		R	T, DO, N, P, S, B, M
Watershed Treatment Model	•	–				S	N, P, S, B
WEPP	•	–	–	–		W	S
WMS	•	•		•	•	I	T, DO, N, P, S, B, Pc, M, A

[a]W – Watershed, R – Receiving water, S – Stormwater, C – Stormwater compliance, and I – Integrated system;
[b]T – Water Temperature, DO – Dissolved Oxygen, N – Nitrogen, P – Phosphorous, S – Sediment, B – Bacteria, Pc – Pesticide, M – Metals, A – Algae

quality interactions involved in a watershed are typically described both spatially and temporally. The watershed area is divided into several grids where all properties, such as management practices, land use, nutrients, crops, soil properties, and slopes, are considered homogenous. The main segments of this model are infiltration, runoff, subsurface drainage, soil erosion, and overland sediment transport.

The most common use of ANSWERS is to predict runoff at a catchment outlet and generate accurate simulations for different surface cover conditions. Runoff predictions are less accurate at low rainfall intensity events when compared to higher intensity events. Another key weakness of ANSWERS is its inability to simulate interflow and groundwater contributions to base flow, snowpack, and snowmelt. Therefore, the model is less suitable for locations where base flow contribution, winter snow accumulation, and snowmelt are high. Moreover, ANSWERS is not adequately formulated to simulate intense single-event storms and has potential numerical problems inherent in its solutions (Borah & Bera, 2003).

2.2 U.S. Environmental Protection Agency Screening Procedures

In most parts of the country, land uses are changing, and it is very important to understand how these changes affect pollution loads and water quality. Some watershed pollution models allow people to factor in various development and agricultural scenarios as well as the changing climate. In this regard, U.S. Environmental Protection Agency (U.S. EPA) screening procedures are an important tool to screen potential practices, narrowing the options down to those that are the most effective and acceptable. As part of the screening procedure, management practices that should be implemented in the critical areas are determined. The criteria for this are based on factors such as pollutant reduction efficiencies, legal requirements, and physical constraints. Once various management options have been identified and screened, the procedures allow users to calculate the effectiveness of the management practices, compare the costs and benefits, and select the final management strategies that will be the most effective in achieving the load reductions needed to meet the goals for the watershed. This screening procedure is an opportunity to identify actions that reduce pollution, restore damaged ecosystems, and protect valuable habitat.

2.3 Regression Method

The Regression Method allows users to model the dependence of one response variable on one or more predictor variables. This method produces objective equations that are easily programmed into comprehensive watershed planning procedures (Moriasi et al., 2007). The equations that relate annual average streamflow to geomorphic, land use, and climatic basin characteristics are easily integrated with geographic information systems (GISs). Moreover, this method offers the opportunity to document the accuracy and uncertainty associated with water yield estimates, including estimation of confidence intervals and information content. Because regression equations document the relationship between climate, geomorphology, and streamflow, they can be used to evaluate the impacts of climate change on water yield. The key advantage of the Regression Method is that the models can be developed to quantify both the mean and variance of annual streamflow; therefore, it provides the complete probability distribution of streamflow for any watershed in a particular region (Vogel, Wilson, & Daly, 1999).

2.4 Simple Method

The Simple Method estimates pollutant loads from the stormwater runoff for urban areas. This method needs an adequate amount of information, including stormwater runoff pollutant concentrations, subwatershed drainage area and impervious cover, and annual precipitation. With the Simple Method, the land use can be broken into specific areas, such as residential, commercial, industrial, and roadway; the model can then calculate annual pollutant loads for each type of land or use more generalized pollutant values for urban runoff. It is also important to note that these values may vary depending on other variables such as the age of development. In addition, the Simple Method only estimates pollutant loads generated during storm events. It does not consider pollutants associated with base flow (Barrett, Walsh, Malina, & Charbeneau, 1998).

2.5 Watershed Analysis Risk Management Framework (WARMF)

The Watershed Analysis Risk Management Framework (WARMF) is a distributed-parameter numerical model used to simulate watershed hydrology and pollutant transport and to develop TMDLs (Chen, Herr, & Weintraub, 2001; McCray, 2005). The output from this model provides information on the water quality of streams, lakes, and many other non-point and point sources of pollution. The WARMF simulates hydrology, including infiltration, runoff, streamflow and routing, evapotranspiration, lake mixing and stratification, and groundwater flow (Tallaksen & Van Lanen, 2004). This model can be applied to the transport of pollutants such as bacterial pollution, dissolved oxygen, acid mine drainage, inputs from septic systems, sediment, mercury, and algae in stratified reservoirs. The key advantages of WARMF are its ability to simulate nonagricultural problems and its physically based formulation. However, WARMF has a few weaknesses. The most important one is that it cannot model deep groundwater aquifers or groundwater quality. Moreover, in the TMDL module, if point-source load reductions are desired, the algorithm will reduce all upstream sources by the same percent.

The WARMF consists of five modules: an engineering module that simulates hydrology and water quality; a consensus module that evaluates management alternatives; a data module that puts the model's input data into graphs and spreadsheets for easy viewing and editing; a knowledge module that stores available information about a watershed; and a TMDL module that guides the user through a process to calculate TMDLs in the watershed. It also provides a decision support system with a user-friendly graphical user interface. The consensus module is useful for presenting the results from the engineering module in a form easily understandable to a varied group of people from technical and non-technical backgrounds, thus making it easier to evaluate management alternatives.

2.6 Agricultural Non-Point Source Pollution Modeling System (AGNPS)

The Agricultural Non-Point Source Pollution Modeling System (AGNPS) is a modeling tool developed by the U.S. Department of Agriculture (USDA) Natural Resource Conservation Service and Agricultural Research Service (ARS) (Bingner & Theurer, 2001; Finn, Scheidt, & Jaromack, 2003). The AGNPS is a single-event model allowing users to evaluate the long-term impact of non-point source pollution from agricultural watersheds as well as the effects of implementing various conservation management alternatives within the watershed. The model operates on a cell basis requiring 22 input parameters. Three input datasets—land cover, digital elevation model, and soil image—are the base data from which AGNPS will extract all required input parameters. The AGNPS provides output under three parameters: hydrology (water runoff), sediment (by particle size class and source of erosion), and chemical (nitrogen, phosphorus, organic carbon, and pesticides) outputs. The next generation of the AGNPS single-event model developed by USDA is called AnnAGNPS. This model is a continuous-simulation, surface-runoff pollutant loading computer model written in standard ANSI Fortran 95 (Yuan et al., 2011). The AnnAGNPS simulates long-term runoff, sediment, and nutrient transport from agricultural watersheds for conservation practices management planning.

2.7 Generalized Watershed Loading Functions (GWLF)

The Generalized Watershed Loading Functions (GWLF) simulates hydrology and loads of sediment, nitrogen, and phosphorus from mixed land uses (Haith, Mandel, & Wu, 1992). Runoff is simulated using curve number and sediment delivery using the universal soil loss equation. The outputs are calculated on a daily basis but reported monthly. The model is fairly simple and requires a low level of expertise. The model limitations are the simplification of stream transport and water quality simulations, lack of routing, and assumption of constant nutrient concentration.

The latest versions of GWLF incorporate new routines for direct simulation of loads from farm animals, a new pathogen load estimation routine, and new detailed urban routines from RUNQUAL for continuous daily simulation of surface runoff and contaminant loads from developed land. This also allows for the evaluation of potential effects of BMPs on contaminated runoff in a given watershed. Three types of BMPs, such as infiltration/retention facilities, vegetated filter strips, and detention basins, can be modeled. An improved version, GWLF-E, is also available. In GWLF-E, the user can specify the extent to which the three BMPs are implemented within any given watershed (Evans & Corradini, 2012). This version also has the ability to simulate the transport and attenuation of pollutant loads from multiple subwatersheds within a larger watershed. GWLF-E has a GIS interface built-in MapWindow called "MapShed".

2.8 Hydrologic Simulation Program FORTRAN (HSPF)/Loading Simulation Program in C++

Hydrologic Simulation Program FORTRAN (HSPF) is one of the most comprehensive watershed/water quality models simulating watershed hydrology, land and soil contaminant runoff, and sediment-chemical interactions because it can simulate a wide range of conventional and toxic organic pollutants (Bicknell, Imhoff, Kittle, Jobes, & Donigian, 2001). The model outputs an hourly time series of simulated parameters. The model uses a lumped approach incorporating rural and urban land segments in subwatersheds; outputs from the subwatersheds are routed through the streams to a larger watershed using a one-dimensional stream and well-mixed reservoir/lake model hydrology and transport. The model is integrated into the U.S. EPA BASINS system. Version 12 includes wetland and shallow water tables, irrigation, simplified snow algorithms, a box model of flow and sediment, and BMP report modules.

The HSPF model uses continuous rainfall and other meteorological records and computes stream flow hydrographs and pollutographs. The three primary modules in HSPF are PERLND, IMPLND, and RCHRES. PERLND deals with the processes on pervious land segments, while IMPLND deals with impervious land segments. They both simulate the water quantity and quality in their respective land segments. The RCHRES module simulates the pollutant transport and fate processes occurring in a stream channel or in a completely mixed lake. WinHSPF provides an easy-to-operate user interface for HSPF (Duda, Kittle, Gray, Hummel, & Dusenbury, 2001).

The Loading Simulation Program in C++ (LSPC) is a watershed model originally developed by rewriting the HSPF code using C++; additional routines were added into the model later (Tetra Tech, Inc. & U.S. EPA, 2002). LSPC also removed some of the limitations in modeling size and model operations, and it has been applied to large, complex watersheds. Output from LSPC has been linked to other model applications, such as EFDC, WASP, and CE-QUAL-W2. The programming architecture used for LSPC allows for easy integration with Microsoft Access and Excel.

2.9 Soil and Water Assessment Tool (SWAT)

The Soil and Water Assessment Tool (SWAT) is a physically based watershed model developed by USDA ARS primarily to predict the effect of land management practices on water, sediment, and agricultural chemical yields in large, complex watersheds with varying soils, land use, and management conditions over long periods of time (Neitsch, Arnold, Kiniry, Srinivasan, & Williams, 2002; Neitsch, Arnold, Kiniry, & Williams, 2011). A continuous simulation model, SWAT typically provides outputs as a daily time series, but can also be used to develop sub-daily outputs.

Model capabilities include pesticide fate and transport simulation and detailed agricultural land planting, tillage, irrigation, fertilization, grazing, and harvesting procedures. The SWAT divides a watershed into subwatersheds. Each subwatershed is connected through a stream channel and further divided into a hydrologic response unit (HRU). The HRU is a unique combination of soil, vegetation type, and land slope in a subwatershed. The SWAT model simulates hydrology, vegetation growth, and management practices at the HRU level. The SWAT model is capable of simulating hydrology using an internal weather generator or user-provided climatic input. The SWAT model is available with GIS interfaces built in ArcGIS, MapWindow, and QGIS (Dile, Daggupati, George, Srinivasan, & Arnold, 2016).

A number of companion tools, such as SWAT Check and SWAT-CUP, are also available to make modeling processes easier. The SWAT Check reads model output from a SWAT project and performs many simple checks to identify potential model problems. The SWAT-CUP is a calibration/uncertainty or sensitivity program interface for the SWAT model. The SWAT-CUP is a public domain program that links various parameter estimation (for calibration) procedures such as SUFI2, PSO, GLUE, ParaSol, and MCMC to the SWAT model. It enables sensitivity analysis, calibration, validation, and uncertainty analysis of SWAT models. The SWAT-CUP has been tested for all procedures before release. The SWAT model also has the option of using a gridded rainfall data that are provided by the RADARs with the help of an extension to SWAT, called NEXRAD-SWAT, that takes the NEXRAD precipitation data estimates and translates them to subbasin-level precipitation inputs.

2.10 Water Erosion Prediction Project (WEPP)

The Water Erosion Prediction Project (WEPP) model is a process-based, distributed-parameter, continuous-simulation erosion prediction model for use on personal computers running Windows 95 and newer (Flanagan & Nearing, 1995; U.S. Forest Service, 2018). The current model version (v2012.8) available for download is applicable to hillslope erosion processes (sheet and rill erosion), as well as simulation of the hydrologic and erosion processes on small watersheds. The download package consists of the WEPP model, WEPP Windows interface, CLIGEN climate generators, documentation, and example data. The WEPP erosion model computes soil loss along a slope and sediment yield at the end of a hillslope. Interrill and rill erosion processes are considered.

Interrill erosion is described as a process of soil detachment by raindrop impact, transport by shallow sheet flow, and sediment delivery to rill channels. Sediment delivery rate to rill flow areas is assumed to be proportional to the product of rainfall intensity and interrill runoff rate. Rill erosion is described as a function of the flow's ability to detach sediment, sediment transport capacity, and the existing sediment load in the flow. The appropriate scales for application are tens of meters for hillslope profiles and up to hundreds of meters for small watersheds. For scales greater than 100 m (328 ft), a watershed representation is necessary to prevent erosion predictions from becoming excessively large.

3.0 URBAN STORMWATER QUALITY MODELS

The following section describes the most common stormwater models that are well suited for a built environment and are used by stormwater professionals. Many of the models described in Section 2.0 are also suitable to be used in an urban/built environment. However, the models listed in this section are specifically suited for a built environment. Several proprietary models, such as MIKE URBAN, CivilStorm, and InfoWorks ICM, are also available, but are not described in this document.

3.1 Program for Predicting Polluting Particle Passage Through Pits, Puddles, and Ponds—Urban Catchment Model (P8-UCM)

The Program for Predicting Polluting Particle Passage through Pits, Puddles, and Ponds—Urban Catchment Model (P8-UCM) is an urban watershed model that can simulate hydrology and water quality components such as total suspended solids (TSS), total phosphorus (TP), total Kjeldahl nitrogen (TKN), copper, lead, zinc, and hydrocarbons (IEP, Inc., 1990). The model handles a variety of BMPs such as swales, buffer strips, detention ponds (dry, wet, and extended), flow splitters, and infiltration basins, pipes, and aquifers. The P8-UCM model performs a continuous water and mass balance on a user-defined system consisting of both pervious and impervious watersheds and runoff storage/treatment areas, particle classes, and water quality components.

The hydrology is based on SCS curve number, Hamon's method for evapotranspiration (ET), and water balance calculation. Water quality from impervious areas are computed using either or both of two techniques: (a) particle accumulation and washoff, and/or (b) fixed-runoff concentration. An exponential washoff relationship similar to that of the SWMM is used to simulate particle buildup and washoff from impervious surfaces. Receiving water processes are limited to devices, ponds, infiltration basins, and shallow channels. Storage area or volume and outflow relations represent flow in ponds. Shallow channel flow is estimated by Manning's equation. Settling and transport of sediments are simulated in the model.

3.2 Source Loading and Management Model (SLAMM)/ WinSLAMM

The Source Loading and Management Model (SLAMM) was developed based on field observations to simulate urban pollutants (Pitt & Voorhees, 2000). The empirical relationships used in SLAMM were derived from U.S. EPA's Nationwide Urban Runoff Program. The SLAMM now also includes a wide variety of source area and outfall control practices (e.g., infiltration practices, wet detention ponds, porous pavement, street cleaning, catch basin cleaning, and grass swales). Beginning with Version 5, SLAMM is Windows-based and thus is called "WinSLAMM". The model performs continuous mass balances for particulate and dissolved pollutants and for runoff volumes.

Runoff is calculated by a method developed by Pitt (1987) for small-storm hydrology, which empirically determines the initial losses and infiltration loss based on experiment data. Runoff is based on rainfall minus initial abstraction, and infiltration is calculated for both impervious and pervious areas. Triangular hydrographs, parameterized by a statistical approach, are used to simulate flow. Exponential buildup and washoff and wind removal functions are used in computing runoff pollutant loadings. The characteristics of the source areas are used to determine pollutant loads in solid and dissolved phases based on an empirical method derived from available field observations. The pollutant removal efficiencies of treatment devices are also estimated based on empirical equations derived from field data.

The WinSLAMM evaluates runoff volume and pollution loading for each source area within each land use for each rainfall event, without lumping impervious areas together. This allows stormwater quality professionals the ability to target the highest loading areas and recommend improvements to reduce runoff volume and pollution loading from those areas. The SLAMM also applies stochastic analysis procedures to more accurately represent uncertainty in model input parameters to better predict the range of outfall conditions (especially pollutant concentrations). The SLAMM applies Monte Carlo sampling procedures to consider the uncertainties in model input values.

3.3 National Stormwater Calculator

U.S. EPA's National Stormwater Calculator is used for estimating the annual amount of rainwater and frequency of runoff from a specific site anywhere in the United States (including Puerto Rico) (Rossman & Bernagros, 2018). Estimates are based on local soil conditions, land cover, and historic rainfall records. A climate change component is included to help explore how climate change impacts may affect the amount of stormwater runoff produced by a site and how it is managed. The tool is designed to be used by anyone interested in reducing runoff from a property, including site developers, landscape architects, urban planners, and homeowners. The calculator accesses several national databases that provide soil, topography, rainfall, and evaporation information for the chosen site.

Moreover, green infrastructure has been included in this estimation. Green infrastructure promotes the natural movement of water, instead of allowing it to wash into streets and down storm drains. The user can choose several green infrastructure practices, such as rain harvesting, rain gardens, green roofs, street planters, infiltration basins, porous pavement, and so on. Additionally, the calculator also allows users to consider how runoff may vary based on historical weather and potential future climate.

3.4 Stormwater Management Model (SWMM)

The Stormwater Management Model (SWMM) was developed by U.S. EPA for the planning, analysis, and design of stormwater, combined, and sanitary drainage systems. The SWMM is used to undertake a single event or long-term simulations of water runoff quantity and quality, primarily, but not limited to, urban areas. It provides an integrated environment for adding input data; running hydrologic, hydraulic, and water quality simulations; and viewing the results in a variety of formats. The SWMM contains a flexible set of hydraulic modeling capabilities used to route runoff and external inflows through the drainage system network of pipes, channels, storage/treatment units, and diversion structures.

In addition, SWMM allows engineers and planners to represent combinations of green infrastructure practices to determine their effectiveness in managing runoff volume (U.S. EPA, 2014). There are a number of versions of the SWMM available, including many commercial versions. These include PCSWMM, XPSWMM, InfoSWMM, and GeoSWMM. These versions incorporate the U.S. EPA-SWMM, but may have additional functionalities including GIS interfaces.

The SWMM5 also incorporated the SWMM Climate Adjustment Tool (SWMM-CAT) that allows future climate change projections to be incorporated into the model. The SWMM-CAT allows users to select monthly climatic data adjustment factors derived from global climate change models run as part of the World Climate Research Programme Coupled Model Intercomparison Project Phase 3. Thus, SWMM-CAT will aid the study of the effect of future changes in climatic conditions. Adjustments can be applied on a monthly basis to air temperature, evaporation rates, and precipitation as well as to the 24-hour design storm at different recurrence intervals.

3.5 Watershed Treatment Model

The Watershed Treatment Model (WTM) represents a simple spreadsheet-based approach that evaluates loads from a wide range of pollutant sources and incorporates watershed treatment options. The model allows the watershed manager to adjust these loads based on the level of effort put forth for implementation. The WTM acts as a starting point from which the watershed manager can evaluate multiple alternatives for watershed treatment, and the model assumptions can be modified as the watershed plan is implemented. The WTM is a simple tool for the rapid assessment and quantification of various watershed treatment options.

The model has three basic components: pollutant sources, treatment options, and future growth. The pollutant sources component of the WTM estimates the load from a watershed without treatment measures in place. The treatment options component estimates the reduction in this uncontrolled load from a wide suite of treatment measures. Finally, the future growth component allows the user to account for future development in the watershed, assuming a given level of treatment for future development.

The WTM allows the watershed manager to assess various programs and sources that are not typically tracked in more complex models. The WTM 2013 is able to track sediment, nutrients, bacteria, and runoff volume on an annual basis. The WTM completes modeling in four steps:

1. calculating pollutant source loads;
2. calculating the benefits of existing practices;
3. calculating the benefits of future practices; and
4. accounting for growth.

4.0 RECEIVING WATER QUALITY MODELS

This section describes the receiving water models that address a receiving waterbody's response to various pollutant loading scenarios. Receiving water models are used to examine the interactions between loadings and response, evaluate loading capacities, and test various loading scenarios.

A differentiating factor among various receiving water models is the type of condition that is handled by the model; namely steady state, quasi-dynamic, and dynamic. A dynamic model allows changes in flow and other conditions at the finest timestep, hourly or even sub-hourly, whereas a steady state model assumes a constant flowrate. Quasi-dynamic allows some variations to evaluate various flow conditions. Another factor is level of details in terms of space and the ability to deal with one, two, and three dimensions. Similar to watershed quality models, receiving water models also differ in their ability to handle various pollutants. Camacho, Zhang, and Chao (2019) present a detailed description of the main capabilities and limitations of 11 widely used receiving water quality models.

4.1 Hydrologic Engineering Center River Analysis System (HEC-RAS)

The USACE's Hydrologic Engineering Center (HEC) developed the River Analysis System (HEC-RAS) to conduct steady and unsteady flow simulation and sediment transport modeling. The HEC-RAS was originally designed to perform one-dimensional hydraulic calculations for a full network of natural and constructed channels. Since the development of Version 5.0, the model also provides two-dimensional modeling capabilities of flow and sediment transfer. The HEC-RAS model provides the capabilities for river flow simulation, floodway encroachment analysis, bridge scour simulation, and channel hydraulic design. The HEC-RAS Version 5.0 also includes a water quality component with the capability to model temperature, nutrients (nutrient profile [NP], dissolved oxygen, carbonaceous biochemical oxygen demand [CBOD], and algae), and arbitrary constituents (simple traces configured by the user).

4.2 QUAL2E

The QUAL2E is a steady state model with the capability of modeling major reactions of nutrient cycles, algal production, benthic and carbonaceous oxygen demand, and atmospheric reaeration. The model is based on the effects of advection, dispersion, dilution, constituent reactions and interactions, and sources and sinks. The model is applicable to sectionally well-mixed, dendritic streams. The modeling reaches are discretized along its longitudinal axis as a series of computation elements, with a maximum allowable number of 50 reaches and 20 computation elements in each reach. Computations are carried out using a finite-difference solution of the advection-dispersion equation with various sources and sink terms.

The QUAL2E models water temperature, chlorophyll-a, organic nitrogen, ammonia, nitrite, nitrate, organic phosphorus, dissolved phosphorus, CBOD, dissolved oxygen, coliforms, an arbitrary nonconservative constituent, and three conservative constituents. The model has a moderate input data requirement that includes headwater conditions for modeled constituents, boundary inputs, temperature, incremental flow, and chemical constants and parameters. The nonpoint-source inputs are assumed to be uniform along the length of a reach and constant in time. The model is provided within the suite of models in U.S. EPA's Better Assessment Science Integrating Point and Non-Point Sources (BASINS). Because QUAL2E is a steady state model, it is not applicable for a dynamic pollutant loading condition. However, it has applicability for TMDL studies in which a point-source-dominated low-flow condition is the main driver. The QUAL2K is a new version of QUAL implemented within Excel.

4.3 Water Quality Analysis Simulation Program (WASP)

The Water Quality Analysis Simulation Program (WASP) is a dynamic multi-dimensional model for simulating contaminant fate and transport in surface waters (Ambrose & Wool, 2009). This model helps users interpret and predict water quality responses to natural phenomena and human-made pollution for various pollution management decisions. The WASP is a dynamic compartment-modeling program for aquatic systems, including both the water column and the underlying benthos. The WASP allows the user to investigate 1-, 2-, or 3-dimensional systems and a variety of pollutant types. The model can simulate processes of advection, dispersion, point and diffuse mass loading, and flux exchange with sediments. The model can also be coupled to external hydrodynamic and sediment transport models.

The state variables that can be modeled include dissolved oxygen, biochemical oxygen demand, nutrients, and metals. The newer versions of WASP contain the sediment diagenesis model linked to the Advanced Eutrophication submodel, which predicted sediment oxygen demand and nutrient fluxes from the underlying sediments. Water quality modules in WASP7 include heat, standard eutrophication, advanced eutrophication, simple toxicants, nonionizing toxicants, organic toxicants, and mercury. The WASP is one of the most widely used water quality models in the United States and throughout the world. Because of the model's ability to handle multiple pollutant types, it has been widely applied in the development of TMDLs. The WASP is also capable of linking with hydrodynamic and watershed models, which allows for multi-year analysis under varying meteorological and environmental conditions.

4.4 TOXMOD

The TOXMOD is a modeling tool developed by the North American Lake Management Society to aid in the assessment of long-term trends and the impact of toxic organic compounds on lakes and impoundments. The model user is required to include variety of input data such as lake depth and surface area, sediment thickness and its area, dissolved organic carbon concentration, time series of flow, and inflow toxicant concentrations. The TOXMOD provides users with a tabular and graphical modeling output of sediments and water toxicant concentration at specified time intervals (U.S. EPA, 1997).

4.5 CE-QUAL-ICM

The CE-QUAL-ICM (known as simply ICM) is a flexible, multidimensional, time-variable water quality model suitable for simulating water quality in lakes and rivers and estuary and coastal waters. The model was developed by the U.S. Army Corps of Engineers (USACE) Engineer Research and Development Center Environmental Laboratory as an adaptable, multidimensional water quality model that can be used for the assessment of variety of waterbodies such as lakes, river, estuaries, and coastal waters (Cerco & Cole, 1995). The model does not simulate hydrodynamics, and information on flow, volume, and diffusion modeling is read into the model using an input file.

The model contains detailed algorithms for water quality kinetics. The ICM can simulate up to 22 state variables including multiple forms of carbon, nitrogen, phosphorus, silica, and dissolved oxygen. The model also includes sediment diagenesis algorithms for predicting sediment oxygen demand and nutrient flux from exchange with the bed. The model structure can be tailored for specific applications, allowing flexibility to add new state variables and processes (U.S. EPA, 1997).

4.6 CE-QUAL-RIV1

The CE-QUAL-RIV1 is a hydro-dynamic, one-dimensional water quality model that can be used to simulate hydrodynamics and water quality in rivers and unstratified lakes (USACE, 1990). The model consists of two modules: RIVH and RIVQ. The RIV1H module solves the unsteady one-dimensional Saint Venant equations to simulate flow, including processes such as advection and dispersion. The output from the RIVH module is used to drive the water quality module, RIVQ. RIVQ can simulate up to 10 water quality variables including temperature, dissolved oxygen, biological oxygen demand, nutrients, metals, and bacteria. The model enables users to assess river systems with multiple hydraulic control structures such as branched river systems and waterway locks and dams.

The input data for CE-QUAL-RIV1 requirements include river geometry (i.e., locations of control structures, river cross-sections, and associated elevations), channel roughness, lateral inflows or withdrawals, and initial and boundary conditions. The model produces output on each water quality aspect depending on a user's requirements and input data provided (U.S. EPA, 1997).

4.7 The River Hydrodynamics Model (RIVMOD-H)

The River Hydrodynamics Model (RIVMOD-H) is a one-dimensional water quality model for gradually or rapidly varying flow in waterbodies. This model can be obtained with the WASP-5 modeling package. The RIVMOD-H model contains a numerically efficient, fully implicit scheme for governing flow equations, allowing this tool to be used for computations with longer timesteps. It can be used to assess flow routing only, or it can be linked with a water quality modeling package such as WASP-5 or SWMM. As input data, RIVMOD-H requires channel morphometry, bed elevations, and initial and boundary conditions to be determined by the user. This model produces time-variable water surface elevations and discharges for unsteady flows at user-specified cross-sections and time intervals (U.S. EPA, 1997).

4.8 Environmental Fluid Dynamics Code (EFDC)

The Environmental Fluid Dynamics Code (EFDC) is a state-of-the-art model that can be used to simulate aquatic systems in one, two, and three dimensions (Tetra Tech, Inc., 2007). It has evolved over the past two decades to become one of the most widely used and technically defensible models for simulating hydrodynamics, water quality (including eutrophication and sediment diagenesis), sediment transport, and contaminant fate and transport in receiving waters. The models solve three-dimensional, vertically hydrostatic, free surface, turbulent averaged equations of motion for a variable-density fluid over a Cartesian or curvilinear grid. The EFDC can simulate wetting and drying of floodplains, mud flats, and tidal marshes. The EFDC can also simulate water and water quality constituent transport in geometrically and dynamically complex surface water environments; the model applications include rivers, lakes, reservoirs, wetlands, estuaries, and coastal ocean regions in support of regulatory permits and requirements, environmental assessments, TMDL development and implementation, and nutrient criteria limit development.

The EFDC computer code solves three-dimensional equations for simulation of fluid flow. As described by the U.S. EPA, the hydrostatic equations of motion used by the model for turbulent flow allow vertical variability. The coordinates used by the model are curvilinear and orthogonal in the horizontal plane. The vertical coordinates are stretched to follow the land surface topography. Free surface displacement in the vertical direction is aligned with the gravitational vector. Turbulent viscosity and diffusivity is related by a second moment turbulence closure scheme to the turbulence intensity and a turbulence length scale. The EFDC solves transport equations for salinity, temperature, suspended cohesive and noncohesive sediment, dissolved and adsorbed contaminants, and a dye tracer. An equation of state relates density to pressure, salinity, temperature, and suspended sediment concentration.

One of the main features of EFDC is its ability to simulate wetting and drying cycles, and it includes a near field mixing zone model that is fully coupled with a far field transport of salinity, temperature, sediment, contaminant, and eutrophication variables. It also contains hydraulic structure representation, vegetative resistance, and Lagrangian particle tracking. The EFDC accepts radiation stress fields from wave refraction-diffraction models, thus allowing the simulation of longshore currents and wave-induced sediment transport.

4.9 CE-QUAL-W2

The CE-QUAL-W2 is a hydrodynamic and water quality model that can be used to simulate lakes, rivers, reservoirs, and river basin systems (Cole & Wells, 2018). This model was originally developed by the USACE Waterways Experiments Station in Vicksburg, Mississippi. The current model release (Version 4.1) enhancements have been developed through research contracts between the USACE and Portland State University under the supervision of Dr. Scott Wells. The model simulation capabilities include eutrophication, sediment diagenesis, an internal dynamic pipe/culvert model, hydraulic structures (spillways, weirs, withdrawal structures), and a dynamic shading algorithm based on topographic and vegetative cover.

The basic eutrophication processes analyzed by the CE-QUAL-W2 model include the relationships between temperature, nutrient, algae, dissolved oxygen, organic matter, and sediment. According to Portland State University, the model capabilities include:

- longitudinal and vertical hydrodynamics and water quality in stratified and non-stratified systems;
- nutrients, dissolved oxygen, and organic matter interactions;

- fish habitat;
- selective withdrawal from stratified reservoir outlets;
- hypolimnetic aeration;
- multiple algae, epiphyton/periphyton, zooplankton, macrophyte, and CBOD;
- sediment diagenesis model and generic water quality groups;
- internal dynamic pipe/culvert model;
- hydraulic structures (weirs, spillways) algorithms, including those for submerged and two-way flow over submerged hydraulic structures; and
- dynamic shading algorithm based on topographic and vegetative cover.

The model is limited by the assumption that the simulated waterbody is well-mixed laterally. In addition, the model does not implement vertical momentum.

4.10 Cornell Mixing Zone Expert System (CORMIX)

The Cornell Mixing Zone Expert System (CORMIX) is a near field hydrodynamic model that predicts plume dilution and geometry resulting from point-source discharge into receiving waters (Jirka, Doneker, & Hinton, 1996). The model simulates near field mixing of discharge plumes with the boundary layer interaction of discharges under steady state conditions. The CORMIX model can simulate plume dynamics resulting from single-port (pipe), multi-port (diffuser), and surface discharges. The CORMIX model contains the following three major subsystems:

- The CORMIX1 is used to predict and analyze environmental impacts of submerged single-port discharges to lakes, rivers, and estuaries.
- The CORMIX2 may be used to predict plume characteristics of submerged multiport discharges.
- The CORMIX3 is used to analyze positively and neutrally buoyant surface discharges to lakes, rivers, and estuaries with a high degree of accuracy.

4.11 BATHTUB

The BATHTUB model is a steady state, one-dimensional model that is designed to simulate water and nutrient mass balances in reservoirs and lakes (Walker, 1996). The model uses a spatially segmented hydraulic network to account for processes such as advective transport, diffusive transport, and nutrient sedimentation. The model uses empirical equations developed using data from reservoirs to simulate the eutrophication processes in lakes and reservoirs. Eutrophication, defined as the nutritional enrichment of waterbodies that leads to an excessive production of organic materials by algae and other aquatic plants, is expressed in terms of total phosphorus, total nitrogen, chlorophyll a, transparency, organic nitrogen, and hypolimnetic oxygen depletion rate. Applications of BATHTUB are limited to steady state evaluations of relations between nutrient loading, transparency and hydrology, and eutrophication responses. Besides short-term responses, responses to variables other than nutrients and effects related to structural modifications cannot be explicitly evaluated.

4.12 Quality Simulation Along River Systems (QUASAR)

Quality Simulation Along River Systems (QUASAR) is a simple dynamic model for nontidal rivers that comprises a set of ordinary differential equations describing the change of flow and concentration of different water quality determinants over a period of time. The model consists of a number of cells in series, with each cell representing a reach. The QUASAR model is applicable only to slowly time-varying flow conditions (Sincock, Wheater, & Whitehead, 2003). This limitation is caused by the use of a single travel-time parameter in both the flow and solute transport models. Under steady-flow conditions, this assumption is valid, although under unsteady conditions, flow travel time represents the kinematic wave velocity (celerity), which is greater than the water velocity and, therefore, is greater than the solute velocity in reaches affected by dead zones (Lees, Camacho, & Whitehead, 1998).

4.13 Curvilinear Hydrodynamics 3D Model (CH3D)

The Curvilinear Hydrodynamics in Three Dimensions (CH3D) model is a three-dimensional model suitable for simulating hydrodynamics, temperature, and salinity in coastal and nearshore waters. The model uses a horizontally boundary-fitted curvilinear grid and a vertically sigma grid to represent the complex bathymetry and shoreline in nearshore waters. The original CH3D model was updated by the USACE Waterways Experiment Station to develop the CH3D-WES model. The CH3D-WES model has been applied to several coastal estuarine systems that include, among others, the Chesapeake Bay, where the model was linked to the CE-QUAL-ICM model to simulate eutrophication kinetics. The original CH3D model was enhanced to include sediment transport (CH3D-SED), water quality (CH3D-WQ3D), light attenuation (CH3D-LA), and seagrass models (CH3D-SAV).

4.14 EUTROMOD

The EUTROMOD model is a collection of spreadsheet tools for predicting the impact of watershed annual average nutrient loadings from watersheds on water quality in lakes (Reckhow, Coffey, Henning, Smith, & Banting, 1992). The model simulates annual average loadings of sediments, nitrogen, and phosphorus using a lumped watershed approach. Lake-wide average water quality conditions are simulated using a set of empirical equations derived from multilake regional datasets. The model has an in-built uncertainly analysis to assess the impact of uncertainty in watershed load predictions on lake instream water quality.

4.15 Visual Plumes

Visual Plumes is a mixing-zone model that simulates surface water jets and plumes in receiving waters (Frick et al., 2003). The model simulates characteristics of single and merged submerged plumes in arbitrary stratified flow conditions. The model can also simulate the fate and transport of bacteria in receiving waters based on temperature, salinity, solar insolation, and water column light absorption. Model applications include mixing-zone studies, TMDL analysis, temperature management plans, antidegradation reports, and other water quality efforts.

4.16 Delft3D

The Delft3D is a multidimensional model that can be used to simulate hydrodynamics, water quality, and contaminant fate and transport in rivers, lakes, and estuaries (Deltares, 2014). The model consists of several integrated modules built around a common interface. The model solves shallow water three-dimensional, vertically hydrostatic, free surface, turbulent averaged equations of motion for a variable-density fluid over a Cartesian or curvilinear grid. The Deltf3D model can simulate wetting and drying of floodplains, mud flats, and tidal marshes. Model applications include rivers, lakes, estuaries, and open ocean in support of support of regulatory permits and requirements, environmental assessments, and hydraulic structure design.

4.17 Tidal Prism Model

The Tidal Prism Model is a steady state model capable of simulating up to 10 water quality variables, including dissolved oxygen and fecal coliform bacteria. The primary strengths and advantages of the Tidal Prism Model are as follows:

- Excellent user documentation and guidance
- Minimal computer storage requirements

- Relatively simple procedures with data requirements that can be satisfied from existing data when site-specific time series data are lacking

The Tidal Prism Model was mainly developed to provide a tool for government agencies for water quality management of small coastal basins (Kuo & Park, 1994). Based on the constituents modeled, it is regarded as a suitable marina mid-range model (Luketina, 1998). The model simulates physical transport using the concept of tidal flushing. It also includes eutrophication processes in water column and benthic sediment, and employs a simple, computationally efficient and accurate solution scheme. This model is applied to smaller estuaries (less than a few kilometers wide) that are internally well mixed (Lakhan, 2003). It is assumed that the water entering the estuary is of oceanic salinity mixing with the fresh river discharge. The Tidal Prism Model is applicable only to waterbodies where tidal forces are predominant with oscillating flow (e.g., an estuary or a tidal river). Therefore, the Tidal Prism Model cannot be applied to waterbodies located on a sound, an open sea, or a lake or reservoir.

5.0 WATER QUALITY COMPLIANCE MODELING

This section describes the models and tools used to meet regulatory compliance needs. Some of them are currently used for a particular state or local jurisdiction, but the models can be used for other areas with or without some changes in the implementation. Some of these models, for example, have a watershed model component. These models provide different opportunities to meet the required water quality compliance targets.

5.1 System for Urban Stormwater Treatment and Analysis Integration (SUSTAIN)

The System for Urban Stormwater Treatment and Analysis Integration (SUSTAIN), developed by U.S. EPA, is a tool that helps stormwater professionals in developing plans for stormwater flow and pollutant controls on a watershed scale (Shoemaker et al., 2009). The tool includes modules for land simulation that are computed using U.S. EPA SWMM, conveyance simulation, BMP simulation, and BMP optimization. The interface built in ArcGIS also includes a BMP siting tool to identify the potential location of possible BMPs and a post-processor to analyze and interpret simulation outputs at different locations. The BMP simulation module simulates flow and pollutant transport for a number of structural BMPs and also includes a BMP cost-estimation module and aggregation of distributed

BMPs. The BMP optimization module implements evolutional optimization techniques, scatter search and non-dominated sorting genetic algorithm-II (NSGA-II), to identify cost-effective BMP placement and selection strategies based on a pre-determined list of feasible sites and applicable BMP types and size ranges (U.S. EPA, 2007).

5.2 Best Management Practice Treatment options for Removal on an Annual basis by those Interested in Nutrients in Stormwater (BMPTRAINS)

This model is used to evaluate Best Management Practice Treatment options for Removal on an Annual basis by those Interested in Nutrients in Stormwater (BMPTRAINS). The calculations in the BMPTRAINS model consist of three major parts:

1. Estimation of annual removal efficiency for nitrogen and phosphorus using a combination of BMPs. The annual runoff volumes in the BMPTRAINS model are computed based on the project meteorological zone location, watershed area, mean annual rainfall depth, percentage input of directly connected impervious area (DCIA), and non-DCIA number.
2. Analysis for user-specified removal efficiency. The BMP is analyzed to see if the specified reduction target is met rather than the removal efficiency found from the difference between the pre- and post-development nutrient loadings.
3. Analysis of individual or multiple BMPs to evaluate their effectiveness.

5.3 Virginia Runoff Reduction Method (VRRM)

The Virginia Runoff Reduction Method (VRRM) was developed by the Commonwealth of Virginia Department of Environmental Quality to ensure that compliance with water quality criteria is achieved in accordance with the Virginia Stormwater Management Program. This modeling tool was developed as two Excel spreadsheets for new development and redevelopment sites. The spreadsheet requires data input on the existing site and proposed site conditions and incorporates different compliance computations in accordance with regulations. As the output, the VRRM spreadsheets provide a summary of land cover statistics including weighted runoff coefficients for the entire site in its developed condition, the pollutant load, and the corresponding treatment volume. Using this tool, the designer can evaluate the effectiveness of different BMPs and their combinations with respect to compliance with water quality and the runoff reduction and pollutant removal regulations (VDCR, 2011).

5.4 Integrated Design and Assessment for Environmental Loadings (IDEAL)

The Integrated Design and Assessment for Environmental Loadings (IDEAL) model was developed to calculate efficiency of pollutant removal within BMPs and treatment systems. It was developed by South Carolina Department of Health and Environmental Control (DHEC) to assist designers and regulators in meeting state and federal requirements. This model can be used for hydrology, sedimentology, and water quality modeling and ties water quality modeling together with physical, chemical, and biological relationships to provide a much more realistic description of reactions. Development of this model was based on literature review, valid scientific principles, and information obtained during site visits at numerous construction locations throughout South Carolina. The IDEAL model can be used to analyze performance of detention/retention ponds, sand filters, and riparian buffers. The performance of each control is modeled using South Carolina Department of Health and Environmental Control-specific conditions such as soils, topography, and climate and compared with removal efficiency (South Carolina DEHC, 2005).

6.0 INTEGRATED MODELING SYSTEMS

This section describes modeling systems that link models, data, and user interface within a single system. They may also include multiple models with linkage between them. Mohamoud and Zhang (2019) present three case studies of the use of integrated modeling systems for development of TMDLs.

6.1 Better Assessment Science Integrating Point and Non-Point Sources (BASINS)

The BASINS model is a multipurpose environmental analysis system designed by U.S. EPA to assist in watershed management and TMDL development by integrating national environmental databases, analysis tools, and watershed and water quality models. The BASINS system includes a variety of databases that are extracted and formatted to facilitate watershed-based analysis and modeling. The databases, obtained from a wide range of federal sources, include base cartographic data, environmental background and monitoring data, and point sources/loading data. The BASINS 4.1 model is built upon the latest stable release of the nonproprietary, open-source MapWindow GIS (U.S. EPA, 2018).

The BASINS model provides user interfaces to easily build watershed models by providing easy download features of base cartographic data such as administrative boundaries, hydrologic boundaries, and major road systems; environmental data such as soil characteristics, land use layers, and stream hydrography; monitoring data that includes data from several existing national water quality, meteorological, stream flow, and groundwater monitoring databases from U.S. EPA, National Oceanic and Atmospheric Administration, National Aeronautics and Space Administration, and United States Geological Survey (USGS); and point-source discharge data. Several watershed and instream models such as HSPF, SWAT, SWMM, PLOAD, WASP, and AQUATOX are integrated into BASINS to allow the user to simulate the behavior of toxic chemicals, conventional pollutants, and nutrients and examine the impacts of these pollutant loadings from point and non-point sources. BASINS also provides a plug-in for GWLF-E. The integrating framework for BASINS is provided by GIS, allowing spatial information as well as analysis output to be displayed as maps, tables, or graphics (U.S. EPA, 2015). The analysis and post-processing tools included in BASINS are GenScn, Climate Assessment Tool, DFLOW, Time Series Functions, and USGS Surface Water Statistics.

6.2 Watershed Modeling System (WMS)

The Watershed Modeling System (WMS) is a comprehensive graphical modeling environment for all phases of watershed hydrology and hydraulics. It was developed by the Environmental Modeling Research Laboratory of Brigham Young University in cooperation with the USACE Waterways Experiment Station. The WMS consists of tools that automate modeling processes, such as automated basin delineation, geometric parameter calculations, and GIS overlay computations (curve number [CN], rainfall depth, roughness coefficients, etc.).

The WMS 7 model supports hydrologic modeling with HEC-1 (HEC-HMS), TR-20, TR-55, Rational Method, NSS, MODRAT, and HSPF. It also supports hydraulic models such as HEC-RAS and CE-QUAL-W2. The WMS model is designed to be modular, enabling the user to select only those modules and hydrologic modeling capabilities that are required. Additional WMS modules can be added at any time. To facilitate data transfer between ArcView GIS and WMS, an extension called "WMSHydro" has been developed. This extension creates a WMS/ArcView "super file", which is a collection of ArcView shapefiles and ASCII grid files. The super file can be exported either from WMS or this extension. The super file also can be imported into WMS or ArcView as themes (coverages) and grids. The WMS includes the following features:

- GIS tools,
- web-based data acquisition tools,

- terrain data import and editing tools,
- automated watershed delineation and hydrologic modeling,
- support for the industry standard hydrologic models,
- step-by-step hydrologic modeling wizard,
- hydraulic modeling and floodplain mapping,
- storm drain modeling,
- two-dimensional (distributed) hydrology,
- integration with Federal Highway Administration (FHWA) hydraulic calculation software, and
- export WMS animations to Google Earth.

7.0 REFERENCES

Ambrose, R. B. & Wool, T. A. (2009). *WASP7 stream transport—Model theory and user's guide: Supplement to Water Quality Analysis Simulation Program (WASP) user documentation* (EPA/600/R-09/100 [NTIS PB2010-104716]). Washington, D.C.: U.S. Environmental Protection Agency.

Barrett, M. E., Walsh, P. M., Malina, J. F., Jr., & Charbeneau, R. J. (1998). Performance of vegetative controls for treating highway runoff. *Journal of Environmental Engineering, 124*(11), 1121–1128.

Bicknell, B. R., Imhoff, J. C., Kittle, J. L., Jr., Jobes, T. H., & Donigian, A. S., Jr. (2001). Hydrological Simulation Program—Fortran (HSPF). User's Manual for Release 12. U.S. EPA National Exposure Research Laboratory, Athens, Georgia, in cooperation with U.S. Geological Survey, Water Resources Division, Reston, VA.

Bingner, R. L., & Theurer, F. D. (2001, March 25–29). AGNPS 98: A Suite of water quality models for watershed use. In *Proceedings of the Sediment: Monitoring, Modeling, and Managing, 7th Federal Interagency Sedimentation Conference*, Reno, Nevada (pp. VII-1–VII-8).

Borah, D. K., & Bera, M. (2003). Watershed-scale hydrologic and nonpoint-source pollution models: Review of mathematical bases. *Transaction of the ASAE, 46*(6), 1553.

Borah, D. K., Ahmadisharaf, E., Padmanabhan, G., Imen, S., & Mohamoud, Y. M. (2019). Watershed Models for Development and Implementation of Total Maximum Daily Loads. *Journal of Hydrologic Engineering, 24*(1).

Camacho, R. A., Zhang, Z., & Chao, X. (2019). Receiving Water Quality Models for TMDL Development and Implementation. *Journal of Hydrologic Engineering, 24*(2).

Cerco, F. C., & Cole, T. (1995). *User's Guide to the CE-QUAL-ICM Three-Dimensional Eutrophication Model* (Version 1.0). Vicksburg, MS: U.S. Army Engineer Waterways Experiment Station.

Chen, C.W., Herr, J., & Weintraub, L. (2001). Watershed Analysis Risk Management Framework (WARMF): Update One—A Decision Support System for Watershed Analysis and Total Maximum Daily Load Calculation, Allocation and Implementation. Publication No. 1005181, Electric Power Research Institute, Palo Alto, CA.

Cole, T., & Wells, S. (2018). *CE-QUAL-W2: A Two-Dimensional, Laterally Averaged, Hydrodynamic and Water Quality Model* (Version 4.1). Portland, OR: Department of Civil and Environmental Engineering, Portland State University.

Deltares. (2014). Delft3D-FLOW Simulation of multi-dimensional hydrodynamic flows and transport phenomena including sediments, User Manual, Hydro-Morphodynamics. Retrieved from https://oss.deltares.nl/documents/183920/185723/Delft3D-FLOW_User_Manual.pdf

Dile, Y. T., Daggupati, P., George, C., Srinivasan, R., & Arnold, J. (2016). Introducing a New Open Source GIS User Interface for the SWAT Model. *Environmental Modelling & Software, 85*(2016), 129–138.

Duda, P. B., Kittle, J. L., Jr., Gray, M. H., Hummel, P. R., & Dusenbury, R. A. (2001). *WinHSPF—An Interactive Windows Interface to HSPF: User's Manual.* Washington, D.C.: U.S. Environmental Protection Agency Office of Water.

Evans, B. M., & Corradini, K. J. (2012). MapShed Version 1.5 Users guide. Penn State Institutes of Energy and the Environment, The Pennsylvania State University. Retrieved from http://www.mapshed.psu.edu/Downloads/MapShedManual.pdf

Finn, M. P., Scheidt, D. J., & Jaromack, G. M. (2003). *User's guide for the agricultural non-point source (AGNPS) pollution model data generator* (USGS Numbered Series No. 2003-130). Reston, VA: United States Geological Survey.

Flanagan, D. C., & Nearing, M. A., Eds. (1995). *WEPP user summary: USDA-Water Erosion Prediction Project* (NSERL Report No. 10). West Lafayette, ID: U.S. Department of Agriculture-Agricultural Research Service National Soil Erosion Research Laboratory.

Foster, G., & Meyer, L. (1972). Transport of soil particles by shallow flow. *Transactions of the ASAE, 15*(1), 99–0102.

Frick, W. E., Roberts, P. J. W., Davis, L. R., Keyes, J., Baumgartner, D. J., & George, K. P. (2003). *Dilution models for effluent discharges.* Washington, D.C.: U.S. Environmental Protection Agency.

Haith, D. A., Mandel, R., & Wu, R. S. (1992). *GWLF: Generalized Watershed Loading Functions user's manual* (Version 2.0). Department of Agricultural & Biological Engineering. Ithaca, NY: Cornell University.

Huggins, L. F., & Monke, E. (1966). The mathematical simulation of the hydrology of small watersheds. *IWRRC Technical Reports*. Paper 2.

IEP, Inc. (1990). *P8 urban catchment model user's manual* (version 1.1). Providence, RI: IEP, Inc.

Jirka, G. H., Doneker, R. L., & Hinton, S. W. (1996). *User's manual for CORMIX: A hydrodynamic mixing zone model and decision support system for pollutant discharges into surface waters*. Washington, D.C.: U.S. Environmental Protection Agency, Office of Science and Technology.

Kuo, A. Y., & Park, K. (1994). *A PC-based tidal prism water quality model for small coastal basins and tidal creeks*. SRAMSOE No. 324, Virginia Institute of Marine Science, Gloucester Point, VA., https://doi.org/10.21220/V5VQ9S

Lackhan, V. C. (2003). *Advances in Coastal Modeling*, Volume 67, First Edition. Elsevier.

Lees, M. J., Camacho, L., & Whitehead, P. (1998). *Extension of the QUASAR river water quality model to incorporate dead-zone mixing*. Hydrology and Earth System Sciences Discussions, European Geosciences Union, 2(2/3), 353–365.

Luketina, D. (1998). Simple Tidal Prism Models Revisited. *Estuarine, Coastal and Shelf Science*, 46(1), 77–84. Retrieved from https://doi.org/10.1006/ecss.1997.0235

McCray, J. E. (2005). Software review: Watershed Analysis Risk Management Framework. Integrated Groundwater Modelling Center, Colorado School of Mines. Retrieved from http://igwmc.mines.edu/software/igwmcsoft/WARMFreview.htm

Mohamoud, Y., & Zhang, Z. (2019). Applications of linked and nonlinked complex models for TMDL development: approaches and challenges. *Journal of Hydrologic Engineering*, 24 (1).

Moriasi, D. N., Arnold, J. G., Van Liew, M. W., Bingner, R. L., Harmel, R. D., & Veith, T. L. (2007). Model evaluation guidelines for systematic quantification of accuracy in watershed simulations. *Transactions of the ASABE*, *50*(3), 885–900.

Neitsch, S. L., Arnold, J. G., Kiniry, J. R., Srinivasan, R., & Williams, J. R. (2002). *Soil and water assessment tool user manual* (Version 2000). (TWRI Report TR-192). College Station, TX: Texas Water Resources Institute.

Neitsch, S. L., Arnold, J. G., Kiniry, J. R., & Williams, J. R. (2011). *Soil and water assessment tool theoretical documentation* (Version 2009). (TWRI Report TR-406). College Station, TX: Texas Water Resources Institute.

Parsons, J. E., Thomas, D. L., & Huffman, R. L. (2004). Agricultural Non-Point Source Water Quality Models, their use and application. *Southern Cooperative Series Bulletin, 398.*

Pitt, R. (1987). *Small Storm Urban Flow and Particulate Washoff Contributions to Outfall Discharges* (Ph.D. dissertation). University of Wisconsin-Madison, Madison, WI.

Pitt, R., & Voorhees, J. (2000). *The Source Loading and Management Model (SLAMM), A Water Quality Management Planning Model for Urban Stormwater Runoff.* Tuscaloosa, AL: University of Alabama, Department of Civil and Environmental Engineering.

Reckhow, K. H., Coffey, S., Henning, M. H., Smith, K., & Banting, R. (1992). *EUTROMOD: technical guidance and spreadsheet models for nutrient loading and lake eutrophication* (draft report); Durham, NC: School of the Environment, Duke University.

Rossman, L. A., & Bernagros, J. T. (2018). *National Stormwater Calculator User's Guide* (Version 1.2.0.1.) (EPA/600/R-13/085e). Revised June 2018; Cincinnati, OH: U.S. Environmental Protection Agency, Office of Research and Development.

Santhi, C., Arnold, J., Williams, J., Hauck, L., & Dugas, W. (2001). Application of a watershed model to evaluate management effects on point and nonpoint source pollution. *Transactions of the ASAE, 44*(6), 1559–1570.

Sharma, A. K., Vezzaro, L. Birch, H., Arnbjerg-Nielsen, K. & Mikkels, P. (2016). Effect of climate change on stormwater runoff characteristics and treatment efficiencies of stormwater retention ponds: A case study from Denmark using TSS and Cu as indicator pollutants. Springerplus, 5(1): 1984.

Shoemaker, L., Riverson, J., Alvi, K., Zhen, J. X., Paul, S., & Rafi, T. (2009). *SUSTAIN—A framework for placement of best management practices in urban watersheds to protect water quality.* Cincinnati, OH: U.S. Environmental Protection Agency National Risk Management Research Laboratory.

Sincock, A. M., Wheater, H. S., & Whitehead, P. G. (2003). Calibration and sensitivity analysis of a river water quality model under unsteady flow conditions. *Journal of Hydrology*, 277(3–4), 214–229.

South Carolina Department of Health and Environmental Control. (2005). *South Carolina DEHC storm water management BMP handbook*. Columbia, SC: South Carolina Department of Health and Environmental Control.

Tallaksen, L. M., & Van Lanen, H. A. (2004). Hydrological drought: processes and estimation methods for streamflow and groundwater. *Developments in Water Science, 48.*

Tetra Tech, Inc., & U.S. Environmental Protection Agency. (2002). The loading simulation program in C++ (LSPC) watershed modeling system user's manual. Fairfax, VA.

Tetra Tech, Inc. (2007). The Environmental Fluid Dynamics Code user manual, U.S. EPA Version 1.01. *Tetra Tech, Inc.* Retrieved from https://www.epa.gov/sites/production/files/2016-01/documents/efdc_user_manual_epa_ver-101.pdf

U.S. Environmental Protection Agency. (1997). *Compendium of Tools for Watershed Assessment and TMDL Development.* EPA-841-B-97-006. Washington, D.C.: U.S. Environmental Protection Agency.

U.S. Environmental Protection Agency. (2007). *SUSTAIN—An EPA BMP Process and Placement tool for urban watersheds.* Retrieved from http://www.epa.gov/NRMRL/wswrd/wq/models/sustain/sustain_paper2007.pdf

U.S. Environmental Protection Agency. (2014). *Storm water management model (SWMM).* Retrieved from https://www.epa.gov/water-research/storm-water-management-model-swmm

U.S. Environmental Protection Agency. (2015). *Better assessment science integrating point and non-point sources (BASINS).* Retrieved from https://www.epa.gov/ceam/better-assessment-science-integrating-point-and-non-point-sources-basins

U.S. Environmental Protection Agency. (2018). BASINS Framework and Features. Retrieved from https://www.epa.gov/ceam/basins-framework-and-features

U.S. Forest Service. (2018). Watershed Erosion Prediction Project (WEPP). *Climate Change Resource Center.* Retrieved from https://www.fs.usda.gov/ccrc/tools/watershed-erosion-prediction-project

Virginia Department of Conservation and Recreation. (2011). *Virginia Runoff Reduction Method Instructions & Documentation.* Retrieved from http://www.dcr.virginia.gov/lr2f.shtml

Vogel, R. M., Wilson, I., & Daly, C. (1999). Regional regression models of annual streamflow for the United States. *Journal of Irrigation and Drainage Engineering, 125*(3), 148–157.

Walker, W. W. (1996). *Simplified procedures for eutrophication assessment and prediction: User manual instruction report.* (W-96-2). Vicksburg, MS: USASCE Waterways Experiment Station. (Updated September 1999).

Yuan, Y., Mehaffey, M. J., Lopez, R. D., Bingner, R. L., Bruins, R., Erickson, C., & Jackson, M. A. (2011). AnnAGNPS model application for nitrogen loading assessment for the future midwest landscape study. *Water, 3*(4), 196–216.

5

Model Selection

Michael Schmidt, P.E., Richard Wagner, P.E., and Steven Wolosoff, BCES

1.0 MODEL SELECTION CONSIDERATIONS

This section will address the selection of modeling tool(s) for watershed and receiving water quality modeling. Selection of modeling tool(s) depends on several factors including model features, budget and time constraints, data availability, required level of accuracy, local and regulatory preferences, and model support. Before a water quality analysis is conducted, several factors need to be addressed. The following subsections present some questions or management decisions to consider for model application.

1.1 What Is the Purpose of Modeling?

The model application should have a clearly designed purpose(s). A simple modeling objective could be the estimation of average annual loads of various water quality constituents from a watershed to satisfy a National Pollutant Discharge Elimination System (NPDES) permit requirement. Examples of more complex objectives could include the evaluation of continuous water quality constituent concentrations in receiving waters and subsequent comparison of results to water quality standards in assessing receiving water impairment. It may also be appropriate to evaluate water quality under both current conditions and projected future conditions, particularly in cases where a currently unimpaired receiving water may be subject to future impacts such as land development. Modeling objectives may also include evaluation to quantify water quality load reductions required for a currently impaired receiving water to meet the water quality standards or a Total Maximum Daily Load (TMDL) requirement.

1.2 What Is the Appropriate Level of Detail?

The required or desired level of detail relates to the modeling purpose as well as available data. In the simple case cited above—evaluating average annual loads for an NPDES permit requirement—the model may need to be discretized to determine the water quality constituent loads at each NPDES outfall. However, watershed models are typically not discretized as much as hydrologic and hydraulic models designed to evaluate flooding and road overtopping impacts. Of course, if the watershed water quality model is an enhancement of an existing hydrologic and hydraulic model, the same discretization would likely be maintained, unless there are reasons to consolidate for the water quality purpose (e.g., run times, focused geographic portion of the model, etc.).

Level of detail should also consider the availability of data. For example, a watershed model and associated receiving water model should be appropriately discretized to establish the watershed tributary area to any associated receiving water monitoring station. The receiving water model should be discretized sufficiently to evaluate the potential impacts of incoming constituent loads, particularly where major loads occur (e.g., near a major water resource recovery facility [WRRF]).

1.3 Which Water Quality Constituents/Processes Are Important?

Generally, the water quality constituents of interest will be those that are causing or potentially causing receiving water impairments. Evaluation of stormwater for NPDES has focused on the following constituents:

- Biochemical oxygen demand (BOD),
- Chemical oxygen demand (COD),
- Total suspended solids (TSS),
- Total dissolved solids (TDS),
- Total Kjeldahl nitrogen (TKN),
- Nitrite (NO_3-) and nitrate (NO_2-),
- Total phosphorus,
- Dissolved phosphorus,
- Cadmium,
- Copper,
- Lead,
- Zinc.

Pathogens and bacteria (fecal coliform, total coliform, *enterococci*, *streptococcus*, *Escherichia coli*) are frequent causes of receiving water impairment. Other constituents of concern can include pesticides and herbicides, toxics (which may include U.S. Environmental Protection Agency [U.S. EPA] priority pollutants), and emerging contaminants (e.g., pharmaceuticals, personal care products, endocrine disrupting compounds, and per- and polyfluoroalkyl substances [PFASs]). The emerging contaminants are a relatively new concern, typically have no water quality standard, and are particularly of concern for drinking water supplies. There is also little information about these constituents relating to content in stormwater (the major sources may be WRRF or other discharges).

Total suspended solids loads to receiving waters can have adverse effects including the destruction of habitat for fish and aquatic life and depletion of storage capacity of lakes and reservoirs. In addition to sediment loads from developed and undeveloped watershed areas, significant loads may be generated by development sites with poor sediment and erosion control and stream scour/erosion as a result of excessive velocities. Simulation of sediment can be related to simulation of other constituents, such as metals and dissolved phosphorus, that may tend to attach to sediment in the water column. Receiving water processes for TSS loads include the settling and scour/erosion/resuspension of bed sediment, which can vary depending upon particle size (e.g., sand vs. silt vs. clay).

Loads of nitrogen and phosphorus can stimulate the growth of algae and other aquatic vegetation, which, in excess, can cause unaesthetic conditions, large swings in diurnal dissolved oxygen, and oxygen depletion when the algae and vegetation die off. In most cases, the water quality standards for nutrients are not a direct result of human health or aquatic life concerns; instead, they reflect nutrient limitations that are required to meet water quality standards for other constituents such as chlorophyll-a content of phytoplankton in the water column. Exceptions could include the un-ionized ammonia standard for aquatic life and nitrate nitrogen limits for drinking water sources.

Receiving water processes for nitrogen and phosphorus loads may include decay from organic to inorganic forms of nitrogen and phosphorus, uptake of inorganic nutrients by algae and vegetation, and settling of particulate nutrients associated with algae or sediment. Benthic release of dissolved nutrients from bed sediments can also be an issue, particularly in lakes that have experienced high loads of nutrients in the past. This release may also be much higher when dissolved oxygen near the interface between the water column and bed sediment is low.

Phytoplankton (free-floating algae), periphyton (attached algae), and macrophytes (aquatic plants) are influenced in receiving waters by the availability of nutrients and light as well as water temperature. As mentioned above, excess levels of nutrients may cause undesirable algae blooms and associated adverse impacts. Macrophytes may include submerged aquatic vegetation (SAV) that are considered desirable for fish and aquatic habitat. Consequently, excessive levels of TSS and phytoplankton in the water column may limit light penetration and inhibit healthy growth of SAV.

Dissolved oxygen concentrations in receiving waters are affected by numerous factors that may include the following sources and sinks:

- Surface reaeration—source
- Sediment oxygen demand (SOD)—sink
- Decay of BOD and ammonia nitrification—sinks
- Algae and macrophyte photosynthesis/respiration—source/sink

In free-flowing streams that are not affected by WRRF discharges, surface reaeration and SOD are the primary source and sink of dissolved oxygen. Water temperature affects dissolved oxygen, as the temperature determines the dissolved oxygen saturation, which, in turn, affects the level of reaeration. Reaeration in free-flowing streams is typically considered a function of flow depth and velocity. In impounded rivers, lakes, and estuaries, the reaeration may be more a function of the wind speed at the surface rather than velocity and depth. Nitrification and BOD decay may be primary factors at and immediately downstream of WRRF discharges or combined sewer overflow (CSO) discharges, causing a dissolved oxygen "sag" downstream until the reaeration source can compensate for the ammonia nitrification and BOD decay. Photosynthesis and respiration are more important in impounded rivers, lakes, and estuaries, in which the residence time is sufficient to promote growth.

1.4 Which Other Model Features May Facilitate the Model Application?

Factors such as familiarity of the modeler/local jurisdictions/regulators with the model, availability of model documentation and user support, data pre-processors and post-processors, and potential interface with other models may be additional considerations. In some cases, states or counties may have prescribed or preferred model tools, and those should be used unless circumstances require use of a different tool. Data pre-processors (to evaluate input datasets for potential errors, generate input datasets using geographic information system [GIS] tools and coverages) and post-processors (plotting model output results and comparing to observed data in the case of model calibration and/or validation) can be useful in more efficient model development and presentation of results to regulators and stakeholders. Model documentation is typically available, but some models also have model support through agencies such as U.S. EPA, online user groups, and archives of questions/answers posed to the user group.

2.0 MODELING PURPOSE

While establishing the objective(s) of the modeling effort, one should consider the current and potential future use of the modeling framework to support water quality decisions. The water quality modeling may be designed to calculate land-based water quality constituent loads only, or may be combined with receiving water modeling to assess the quality (loads and concentrations) of the receiving water relative to applicable water quality standards, criteria, or goals.

Consequently, potential objectives can range from a simple estimation of average annual or seasonal pollutant loads at the defined study area outlet to a continuous simulation of concentrations at various points within the study area. Typically, the water quality evaluation has been initiated by existing monitoring data that indicate that water quality may be impaired or projections (e.g., anticipated future land use changes) that indicate potential water quality impairment in the future. Seasonal or annual load calculations can be used to determine whether existing or future loads are consistent with an established total maximum daily load for a receiving water. In contrast, continuous simulation of land-based loads and receiving water concentrations can determine whether receiving water standards, criteria, or goals are met. These may include standards, criteria, or goals for the land-based model constituents (e.g., BOD, nutrients) and receiving water constituents that are a byproduct of other constituents (e.g., nutrient impact on algae concentrations, BOD impacts on dissolved oxygen concentrations).

Generally, the initial modeling objective is to develop a tool that provides a valid representation of current water quality conditions, but it may ultimately include use of the tool to evaluate potential future conditions and alternative management measures to reduce pollutant loads and improve water quality. Future conditions may include changes in land use (e.g., conversion of undeveloped or agricultural land to urban development) and wastewater discharges (e.g., increased WRRF discharge with population changes, modification of collection systems to reduce CSO and/or sanitary sewer overflow [SSO] discharges). In the case of land use impacts, the analysis may consider several stormwater management alternatives that include the existing stormwater treatment requirements for new development, potential alternative stormwater treatment requirements for new development, modifications to future land use plans/zoning, or the addition of "retrofit" stormwater treatment facilities to areas that developed before stormwater treatment was required.

3.0 APPROPRIATE LEVELS OF DETAIL

Table 5.1 presents examples of various levels of complexity in water quality modeling. Considerations presented in the table include hydrology/hydraulics, land-based water quality simulation, evaluation of stormwater best management practices (BMPs), evaluation of water quality processes in receiving waters, and pre- and post-processing for model development and analysis of results.

Watershed hydrology/hydraulics can range from a simple estimation of average annual or seasonal flows at the study area outlet based on runoff coefficients to a continuous simulation of flows generated from the land surface and routed through the study area conveyance network. Continuous land surface flows from pervious land areas are calculated using daily or sub-daily (e.g., hourly) rainfall and accepted methods for converting rainfall to surface runoff, and potentially also tracking the fate of infiltrated stormwater with respect to evapotranspiration, base flow, and deep aquifer recharge. Runoff from impervious land areas includes all rainfall except what is captured by available surface depression storage (which is depleted by evaporation between storms).

TABLE 5.1 Levels of complexity in water quality modeling.

Modeling Aspect	Level of Complexity: Low	Moderate	High
Hydrology/ hydraulics	Runoff coefficient, curve number, unit hydrograph routing	Empirical water storage accounting, rating curve for hydraulic conveyance	Continuous, nonlinear reservoir overland flow, channel and reservoir hydraulics
Land-based water quality	EMC, export coefficients	Buildup and washoff	Explicit sediment and adsorption kinetics
Stormwater BMPs	Simple pollutant removal factors	Nonlinear pollutant removal functions	Optimization of BMP site and type selection, probabilistic analysis
Receiving water processes	Simple load reduction	First-order decay rate	Sediment scour/ deposition, first-order decay rate, nutrient cycling
Model pre- and post-processing	Spreadsheet, simplified graphical user interface (GUI)	GIS data import, time series data input/output	GIS import, parameter changes in text editor, complicated GUIs

Corresponding surface runoff constituent concentration and load data can range from simple event mean concentration (EMC) values assigned to runoff from various land-use types, to surface buildup and washoff parameterization for various land uses, to explicit simulation of surface and subsurface processes. Values for EMC can be estimated based on local or nationwide stormwater runoff monitoring data, such as that found in the National Stormwater Quality Database. The same data may be used to establish reasonable buildup and washoff coefficients such that the modeled mean and variability of runoff event concentrations are representative of the local or nationwide data. The buildup and washoff approach is more suitable in cases where the variability of receiving water constituent concentrations is important (e.g., evaluating against a water quality standard expressed as a concentration). Explicit modeling of surface and subsurface processes may be appropriate for cases such as evaluation of sediment nutrient and pesticide loads from agricultural lands that are affected by tillage practices and the application methods and rates for pesticides and fertilizers.

As mentioned earlier, some models allow for the explicit modeling of groundwater flow, whereas other models may account for groundwater quantities by establishing groundwater flow as an input time series. Few models explicitly simulate groundwater quality, which is typically assigned a concentration that is either constant or can vary monthly. Representative concentrations may be determined by reviewing observed instream concentrations for samples taken during dry weather conditions in areas not affected by wastewater discharges. Again, detailed models considering nutrient and pesticide loads from agricultural lands may include explicit simulation of subsurface concentrations.

Most land-based water quality models include the capability of accounting for surface runoff load reductions caused by stormwater BMP facilities. Whether or not the effect of BMPs needs to be considered under current watershed conditions may depend on the age of the urban development with respect to the implementation of stormwater treatment regulations in that jurisdiction. When evaluating future conditions, accounting for BMPs may be important, particularly if substantial urban development is anticipated and alternative stormwater treatment regulation will be considered. Level of detail for BMP consideration can range from simple assumed removal efficiencies applied to runoff from areas served by BMPs to explicit modeling of BMP facilities that account for the runoff quantity captured by BMPs and the calculation of removal based on factors such as residence time in the BMP facility.

Watershed routing through "upland" (i.e., non-tidal) areas may range from a simple summation of inflows to the receiving waterbody to a more complex dynamic routing of the flows and loads. Dynamic routing can account for phenomena such as backflow/tailwater effects, branching networks, hysteresis, and operations of hydraulic structures. Moderate levels of complexity may include flow routing based on rating curves (e.g., stage–discharge relationships) that are considered applicable under all conditions.

Receiving water simulations of lakes, larger rivers, and tidal rivers and estuaries may consider two-dimensional or three-dimensional modeling. Considerations may include vertical stratification and incomplete horizontal mixing of incoming flows and loads (e.g., discharge "hugs the shore" in a larger river).

Water quality processes for water quality constituents represented in the receiving water model may vary from a simple load delivery estimate based on a corresponding estimate of travel time through the study area, to a simple loss rate based on first-order decay, to explicit simulation of various loss and transformation processes.

The desired level of detail may also depend on funding and schedule constraints and available data for model calibration and validation. It may be appropriate to do a relatively simple model first to answer questions such as relative sources of pollution and anticipated benefits of alterative management measures, and then do a more detailed analysis later. This approach could also best allocate time and resources to collect data to support the more detailed analysis.

The following sections will feature information on a number of models that are capable of water quality simulations based on calculations of land-based flows and loads, receiving water flows, loads and concentrations, or both. The selected models are believed to represent a range of applications with respect to land use, temporal and spatial scales, water quality constituents, and associated land-based and receiving water processes.

4.0 DETERMINATION OF TEMPORAL AND SPATIAL SCALES

In a modeling analysis, the temporal scale refers to both the time over which water quality changes occur in the environment and the associated modeling time step. At its simplest state, a water quality modeling analysis may include the calculation of seasonal or average annual water quality constituent loads, essentially having a seasonal or annual time step, for an entire study area or watershed. For continuous simulation, the simulation time period for land-based loads in many cases may be a particular "average year" based on comparison of a year's rainfall patterns to long-term average patterns, or it may be a series of consecutive years that include an average year, a wet year, and a dry year based on annual rainfall to test water quality impacts over a range of conditions. In some cases, it may even be appropriate to simulate a decade or more to see how long it will take for a receiving water to recover from a "legacy" load (i.e., high levels of nutrients in lake sediments that will be steadily released from the sediments even after the nutrient load to the lake has been reduced).

Model time steps for routing through a watershed conveyance network, stream, lake, or estuary will generally be at a smaller time scale than the model time step for land-based flow and load calculation. Land-based continuous simulation of flows and loads will typically be based on daily or sub-daily (e.g., hourly) time steps, depending on how the hydrologic method is applied by the model. When the land-based flows are routed through the watershed, simple kinematic wave routing or storage-discharge routing can use similar time steps to watershed load calculation (hourly to sub-hourly), whereas dynamic wave or hydrodynamic routing will require time steps of minutes or even seconds.

Tables 5.2 and 5.3 provide summary information on spatial and temporal scale for selected models. Table 5.2 focuses on land-based models that generate flows and loads from the study area, whereas Table 5.3 focuses

TABLE 5.2 Temporal and spatial extents for land-based water quality simulation models.

Land-Based Model	Spatial and Temporal Extent			
	Level of Detail	Spatial Extent	Land Cover	Temporal Scale
AGNPS	Detailed	Watershed	Agricultural	Continuous
BMPTRAINS	Screening	Site	Urban	Annual
GLEAMS	Screening	Site	Agricultural	Continuous
GWLF	Detailed	Watershed	Agricultural, rural, urban	Continuous
HSPF	Detailed	Watershed	Agricultural, rural, urban	Continuous
OPUS	Screening	Site	Agricultural	Continuous
P8-UCM	Screening	Site	Urban	Continuous, event
SWAT	Detailed	Watershed	Agricultural	Continuous
SWMM	Detailed	Site, Watershed	Rural, urban	Event, continuous
VRRM	Screening	Site	Urban	Annual
WARMF	Detailed	Watershed	Agricultural, rural, urban	Continuous

TABLE 5.3 Receiving water model temporal and spatial extents for water quality simulation.

Receiving Water Model	Type of Waterbody	Dimensionality	Temporal Scale
BATHTUB	Lake	1-Dimensional	Seasonal, annual
CE-QUAL-ICM	Stream, lake, estuary	1-, 2-, 3-Dimensional	Continuous
CE-QUAL-W2	Stream, lake, estuary	2-Dimensional (laterally averaged)	Continuous
EFDC	Stream, lake, estuary	1-, 2-, 3-Dimensional	Continuous
EPDRiv1	Stream	1-Dimensional	Continuous
HSPF	Stream, lake	1-Dimensional	Continuous
QUAL2K	Stream	1-Dimensional	Steady flow event
SWMM	Stream, lake	1-Dimensional	Event, continuous
WARMF	Stream, lake	1-Dimensional	Continuous
WASP	Stream, lake, estuary	1-, 2-, 3-Dimensional	Continuous

on receiving water models. In some cases, a model has both land-based and receiving water simulation capabilities, and those models are included in both tables. The land-based models featured here generally fall within the watershed, urban stormwater, and water quality compliance models described in Chapter 4. For both land-based and receiving water model tables, the selected models represent a subset of all available models that are nonproprietary and exhibit a variety of complexity and model features for comparison.

Spatial scale refers to the suitable study area size and discretization for model application. Some land-based models may be suitable to small-scale modeling (e.g., development sites and small watersheds) with limited conveyance system flow routing and pollutant transformations, whereas others are more suited to large watersheds. The discretization of the study area and receiving waters may be determined by factors such as the locations of flow and water quality monitoring stations, locations of major pollution sources, and other factors. Most of today's models have no limitations on the number of land-based subcatchments or receiving water reaches/segments.

Table 5.2 summarizes land-based model characteristics, including the following:

- **Level of Results.** The models were evaluated as "screening" or "detailed". In general, "screening" suggests that the model results are suitable for determining relative differences between alternative watershed conditions or management measures, but it may not provide accurate absolute calculation of watershed water quality constituent loads. Those classified as "detailed" are better suited to provide results that are consistent with measured flows and loads, provided that there are sufficient available data for model calibration and validation.
- **Spatial Extent.** The models were evaluated as "site" or "watershed". In general, "site" suggests that the model is appropriate for evaluating an urban development or doing a "field-scale" analysis in the case of most agricultural models in the table. Those classified as "watershed" are better suited to provide results for a larger study area, incorporating the capability to simulate multiple sub-areas and combine those loads to establish the total study area load. Most of these models also have the receiving water capability to account for load reductions and transformations through the watershed conveyance network.
- **Land Cover.** The models were evaluated as "agricultural", "rural", "urban", or a combination of those. The agricultural models have detailed simulation specific to agricultural practices such as

fertilization and pesticide application, and more explicit simulation of nutrients and pesticides in surface runoff and subsurface water. Those models that are classified strictly as "urban"—the Program for Predicting Polluting Particle Passage through Pits, Puddles, and Ponds–Urban Catchment Model (P8-UCM), Virginia Runoff Reduction Method (VRRM), Best Management Practice Treatment Options for Removal on an Annual Basis by those Interested in Nutrients in Stormwater (BMPTRAINS)—are geared toward the site level and are oriented toward evaluation of runoff loads for pre-development and post-development conditions, and BMP facility effectiveness. Those classified as both rural and urban are more appropriate for watersheds having a mix of urban areas and undeveloped areas.

- **Temporal Scale.** The models were evaluated as "annual", "continuous", "event", or a combination of those. The annual models (VRRM, BMPTRAINS) reflect the simple approach of estimating annual runoff from annual rainfall and runoff coefficients based on impervious cover and soil types for pervious land within the site. Most models are designated as continuous, requiring a time series of rainfall to drive the model and generating results on a daily or sub-daily basis. Generally, land-based models are not applied on an event basis.

Table 5.3 summarizes receiving water model characteristics, including the following:

- **Type of Waterbody.** The models were evaluated as "stream", "lake", "estuary", or a combination of those waterbody types. In general, the estuary designation was applied only to those models that can account for tidal hydrodynamics and associated water quality constituent exchange at the tidal boundary. (The Stormwater Management Model [SWMM] model can handle tidal flow reversals, but it does not have the explicit capability of defining a boundary concentration.) Most of the other models are designated as both stream and lake; however, use of some models for lake simulation may not be optimal based on vertical stratification (see discussion on dimensionality). The BATHTUB model is an empirical model that is designed strictly for lakes and impoundments, and QUAL2K and the CE-QUAL-RIV1 variant EPDRiv1 are specifically designed for rivers.
- **Dimensionality.** Most of the models in the table are evaluated as one-dimensional, which means a model reach or segment is completely mixed (i.e., no vertical or lateral differences in concentration). The CE-QUAL-W2 model is specifically suited to evaluating receiving

water quality with vertical stratification, but it does not account for lateral variability of concentrations. The WASP, CE-QUAL-ICM, and Environmental Fluid Dynamics Code (EFDC) models have the flexibility to model water quality in one-, two-, or three-dimensional fashion.

- **Temporal Scale.** Most of the models are designated as "continuous" and would have time steps, as discussed earlier. The BATHTUB model is designed to evaluate nutrient and algae concentrations on a seasonal or annual basis. The QUAL2K model is designed to evaluate concentrations under steady flow conditions, such as a WRRF discharge under critical low-flow/high-water temperature conditions, which was considered a "steady flow event" simulation in the table. The SWMM also has an event designation, considering that it may be used to generate event flows from sources such as CSOs or SSOs for a designed storm event.

5.0 REQUIRED MODEL FEATURES AND CAPABILITIES

Tables 5.4 through 5.6 provide summary information on water quality simulation for the featured land-based models and the methods used to calculate flows and loads for surface runoff and groundwater (base flow). Table 5.4 identifies the water quality constituents that can be simulated by the featured models. Tables 5.5 and 5.6 discuss the methods used by the models to simulate the identified potential constituents of interest in surface water runoff and groundwater.

Tables 5.7 through 5.9 provide similar data for the featured receiving water models. Table 5.7 identifies the water quality constituents that can be simulated by the featured models, and the latter tables focus on receiving water processes related to sediment in the water column and bed sediments (Table 5.8) and reaeration (Table 5.9).

The following subsections compare and contrast the featured models for the most commonly modeled water quality constituents. The discussion includes both the land-based models (sources for receiving waters) and the processes within the receiving waters.

5.1 Water Temperature

Water temperature is a critical element for event or continuous water quality modeling in receiving waters because it affects the rate of kinetic processes. Of the featured land-based models, only a few of the more complex models—Hydrologic Simulation Program FORTRAN (HSPF), Soil and

TABLE 5.4 Modeled water quality constituents for land-based water quality simulation models.

Land-Based Model	Water Temperature	Dissolved Oxygen	BOD	Sediment	Nitrogen	Phosphorus
AGNPS	No	No	No	Yes	Yes	Yes
BMPTRAINS	No	No	No	No	No	No
GLEAMS	No	No	No	Yes	Yes	Yes
GWLF	No	No	No	Yes	Yes	Yes
HSPF	Yes	Yes	Yes	Yes	Yes[a]	Yes[a]
OPUS	No	No	No	Yes	Yes	Yes
P8-UCM	No	No	No	Yes	Yes[b]	Yes[c]
SWAT	Yes	Yes	Yes	Yes	Yes	Yes
SWMM	No	No	Yes	Yes[d]	Yes[d]	Yes[d]
VRRM	No	No	No	No	Yes	Yes
WARMF	Yes	Yes	Yes	Yes	Yes	Yes

TABLE 5.4 Modeled water quality constituents for land-based water quality simulation models. (*Continued*)

Land-Based Model	Water Quality Constituent					
	Bacteria	Pesticides	Metals	Alkalinity	pH	Other
AGNPS	Yes	Yes	No	No	No	No
BMPTRAINS	No	No	No	No	No	No
GLEAMS	No	Yes	No	No	No	No
GWLF	Yes	No	No	No	No	No
HSPF	Yes	Yes	Yes[d]	Yes	Yes	Yes
OPUS	No	Yes	No	No	No	No
P8-UCM	No	No	Yes	No	No	Yes
SWAT	Yes	Yes	Yes	Yes	Yes	Yes
SWMM	Yes[d]	Yes[d]	Yes[d]	No	No	Yes
VRRM	No	No	No	No	No	No
WARMF	Yes	Yes	Yes	No	No	Yes[e]

[a]May be modeled explicitly or as general pollutant
[b]Documentation says TKN, not total N
[c]Documentation says Total P (not species)
[d]General pollutant
[e]Documented in text

TABLE 5.5 Hydrologic and water quality processes for land-based water quality simulation models: surface runoff.

Land-Based Model	Surface Runoff Flow	Surface Runoff Quality
AGNPS	Models using curve number (daily precipitation).	Models using revised universal soil loss equation (RUSLE); sediment transported nutrients and pesticides.
BMPTRAINS	Models mean annual runoff as function of directly connected impervious area (DCIA) and non-DCIA curve number.	Models using total nitrogen and total phosphorus EMCs by land-cover category.
GLEAMS	Models using curve number (daily precipitation).	Models using modified universal soil loss equation (USLE).
GWLF	Models using curve number for rural and urban areas (daily precipitation).	Rural: Models using USLE and sediment delivery ratio, dissolved nitrogen and phosphorus coefficients, and sediment nitrogen and phosphorus coefficients. Urban: Models buildup/washoff (assumed to be in solid phase).
HSPF	Pervious: Models based on Stanford Watershed Model. Impervious: Models 100% runoff after depression storage is filled (hourly or sub-hourly precipitation).	Pervious: Models sediment detachment/attachment; other constituents buildup and washoff, potency factor, or both. Impervious: Models buildup/washoff, potency factor, or both.
OPUS	Models using curve number or breakpoint infiltration.	Models using modified universal soil loss equation (USLE) or more detailed option; up to 10 soluble or adsorbed pesticides; nutrient washoff equilibrated with sediment in shallow mixing layer.
P8-UCM	Pervious: Models using curve number. Impervious: Models 100% impervious runoff after depression storage filled (hourly precipitation).	Pervious: Models exponential washoff. Impervious: Models buildup/washoff, EMC, or both.
SWAT	Pervious: models using curve number or Green-Ampt infiltration. Urban areas: Models curve number.	Pervious: models using modified USLE; nutrients, pesticides based on content in surface layer, sediment load. Urban areas: models United States Geological Survey regression equations or buildup-washoff.
SWMM	Pervious: Models using Horton or Green-Ampt infiltration for areas. Impervious: models 100% impervious runoff after depression storage filled (hourly or sub-hourly precipitation); can route impervious runoff to pervious area.	Models using EMC or buildup/washoff for all constituents.
VRRM	Models using runoff coefficients for forest/open, managed turf, and impervious cover (annual rainfall).	Models using total nitrogen and total phosphorus EMCs by land-cover category.
WARMF	Pervious: models explicit simulation of surface storage, infiltration and runoff of excess. Impervious: Models immediate runoff.	Pervious: models sediment detachment/attachment, other constituents' buildup and then washoff based on modeled surface concentration.

TABLE 5.6 Modeled hydrologic and water quality processes for land-based water quality simulation models: groundwater.

Land-Based Model	Groundwater Flow	Groundwater Quality
AGNPS	Models explicitly	Models explicitly
BMPTRAINS	Does not model	Does not model
GLEAMS	Models explicitly	Models explicitly
GWLF	Estimates as difference between infiltration and evapotranspiration	Models based on assigned total nitrogen and total phosphorus concentrations (dissolved)
HSPF	Models based on Stanford Watershed Model—may include deep recharge loss from watershed	Models explicitly (nutrients) or based on assigned annual/ monthly concentrations
OPUS	Models explicitly	Models explicitly
P8-UCM	Estimates as difference between infiltration and evapotranspiration	Models, but it is unclear if groundwater carries a water quality constituent load
SWAT	Models explicitly—may include deep recharge loss from watershed	Models explicitly for nitrate and pesticides; concentration can be assigned for sediment and soluble phosphorus
SWMM	Models explicitly—may include deep recharge loss from watershed—or assigns as dry weather flow	Models based on assigned concentrations (constant)
VRRM	Does not model	Does not model
WARMF	Models explicitly	Models explicitly

Water Assessment Tool (SWAT), Watershed Analysis Risk Management Framework (WARMF)—have the capability of simulating water temperature. Temperature (heat) associated with land-based loads may not be critical in large receiving waters where receiving water processes dominate the heat budget, but it may be important in smaller urban streams subject to higher temperature runoff from impervious areas that absorb heat in the summer months, resulting in higher runoff and instream temperatures. Most of the featured receiving water models can simulate water temperature as a function of other meteorological data such as air temperature, solar radiation, wind speed, and cloud cover. Alternatively, time series of water temperature may be input to some models based on available monitoring data. Neither SWMM (which has limited receiving water kinetics that are not related to water temperature in the model) nor BATHTUB (which calculates seasonal or annual results based on empirical lake data) model water temperature.

TABLE 5.7 Modeled water quality constituents for receiving water models.

Receiving Water Model	Water Quality Constituent							
	Water Temperature	Dissolved Oxygen	Biochemical Oxygen Demand	Sediment	Nitrogen	Phosphorus	Bacteria	Pesticides
BATHTUB	No	No	No	No	Yes	Yes	No	No
CE-QUAL-ICM	Yes	Yes	Yes	Yes	Yes	Yes	No	No
CE-QUAL-W2	Yes	Yes	Yes	Yes[a]	Yes	Yes	Yes	Yes
EFDC	Yes	Yes	Yes	Yes	Yes	Yes	Yes	No
EPDRiv1	Yes	Yes	Yes[b]	No	Yes	Yes	Yes	No
HSPF	Yes	Yes	Yes	Yes[a]	Yes	Yes	Yes[a]	Yes
QUAL2K	Yes	Yes	Yes[b]	Yes	Yes	Yes	Yes	Yes[c]
SWMM	No	No	Yes[c]	Yes[c]	Yes[c]	Yes[c]	Yes[c]	Yes[c]
WARMF	Yes	Yes	Yes	Yes	Yes	Yes	Yes	Yes
WASP	Yes	Yes	Yes[b]	Yes[a]	Yes	Yes	Yes	No

TABLE 5.7 Modeled water quality constituents for receiving water models. (*Continued*)

Receiving Water Model	Water Quality Constituent								
	Phytoplankton	Periphyton	Macrophytes	Zooplankton	Metals	Alkalinity	pH	Total dissolved Solids/Salinity	Other Constituents
BATHTUB	Yes	No	No	No	No	No	No	No	No
CE-QUAL-ICM	Yes[d]	No	No	No	No	No	No	Yes	No
CE-QUAL-W2	Yes[d]	Yes[d]	Yes	Yes	Yes	Yes	Yes	Yes	Iron
EFDC	Yes[d]	Yes	Yes	No	No	No	No	Yes	Silica
EPDRiv1	Yes	No	Yes	No	Yes	No	Yes	No	No
HSPF	Yes	Yes	No	Yes	Yes	Yes	No	Yes	No
QUAL2K	Yes	Yes	No	No	No	No	No	No	No
SWMM	No	No	No	No	Yes[c]	Yes[c]	No	Yes[c]	No
WARMF	Yes[d]	Yes	No	No	Yes	Yes	Yes[e]	Yes	Yes[f]
WASP	Yes[d]	Yes	No	No	Yes	Yes	Yes	Yes	Silica

[a]Multiple particle sizes
[b]Multiple BOD species with varying decay rates
[c]General pollutant
[d]Multiple types
[e]Model's alkalinity and total inorganic carbon, which are basis for pH
[f]Documented in text

TABLE 5.8 Modeled water quality processes for receiving water models: sediments.

Receiving Water Model	Streambank Erosion/ Bottom Scour/Deposition	Sediment Exchange With Water Column
BATHTUB	Does not model	Does not model
CE-QUAL-ICM	Inorganic suspended solids settling rate	Sediment diagenesis calculation
CE-QUAL-W2	Inorganic suspended solids settling rate	SOD and nutrient flux rates
EFDC	Inorganic suspended solids settling or resuspension based on modeled shear stress and critical shear stress	Sediment diagenesis calculation
EPDRiv1	Does not model	SOD and nutrient flux rates
HSPF	Streambank erosion/scour based on scouring rate and critical velocity for scour (silt, clay); several options for suspended sand	SOD and nutrient flux rates
QUAL2K	Does not model	Does not model
SWMM	Could estimate suspended solids settling as first-order decay	Does not model
WARMF	Streambank erosion based on empirical factors and stream velocity; scour based on scouring rate and critical velocity for scour; inorganic suspended solids settling rate for silt and clay; bed load capacity calculation for sand deposition	SOD rate
WASP	Models inorganic suspended solids settling and scour rates	Models SOD and nutrient flux rates or sediment diagenesis calculation

5.2 Dissolved Oxygen

Dissolved oxygen simulation is contingent upon many of the other water quality constituents previously discussed such as water temperature, decay of CBOD and/or organic matter, nitrification of ammonia nitrogen, algal growth and respiration, and SOD. Another process affecting modeled dissolved oxygen is reaeration, which is discussed below.

Generally, many of the featured land-based models do not explicitly account for the dissolved oxygen associated with surface runoff or groundwater inflow. The SWMM can model dissolved oxygen concentration, but only as an EMC. In the HSPF model, surface runoff dissolved oxygen is assumed to be at saturation and, therefore, is dependent on the modeled surface runoff temperature. Groundwater dissolved oxygen is assigned in HSPF as either a constant value or monthly values. The SWAT model calculates

TABLE 5.9 Modeled water quality processes for receiving water models: reaeration.

Receiving Water Model	Reaeration Rate Options			
	User-Defined	Hydraulic (Velocity, Depth, Slope)	Wind	Weir/Spillway
BATHTUB	No	No	No	No
CE-QUAL-ICM	Yes[a]	No	No	No
CE-QUAL-W2	Yes[b]	Yes	Yes[c]	Yes
EFDC	No	Yes	Yes	No
EPDRiv1	Yes[a,b]	Yes[d]	Yes[d]	Yes
HSPF	Yes[b]	Yes	Yes[c]	No
QUAL2K	No	Yes	Yes	Yes
SWMM	No	No	No	No
WARMF	No	Yes	Yes	No
WASP	Yes[a]	Yes	Yes	Yes

[a]Constant value
[b]Formula related to depth, velocity, slope
[c]Used for lakes only
[d]Larger calculated value based on hydraulics or wind speed is used as the modeled rate

surface runoff dissolved oxygen as saturation (assuming rainfall is at saturation) minus the impact of CBOD decay in the surface runoff, which reflects the CBOD content of the runoff and the overland flow travel time. It is not clear if SWAT groundwater flow includes any dissolved oxygen.

Receiving water models often offer several different methods for evaluating reaeration that typically relate to flow depth and velocity. In lakes that are deep and have very low velocities, wind at the surface may control the rate of reaeration. Table 5.9 summarizes the options for the featured models and indicates whether the models can accept user-specified constant rates or parameter values used to calculate reaeration; internal calculation methods using modeled depth, width, and slope; calculation methods for evaluating wind reaeration; and calculation methods for calculating reaeration caused by flow over a spillway.

Most of the featured models include one or more internal options for calculating reaeration based on simulated hydraulics (SWMM and BATHTUB do not model receiving water dissolved oxygen and CE-QUAL-ICM is limited to user-defined values). The WARMF model appears to be limited to reaeration based on hydraulic simulation (streams) or wind reaeration (lakes). The HSPF model does not explicitly account for dam reaeration; although, the model does include special actions routines that could be programmed to account for dam reaeration using similar algorithms as the other models. Both the HSPF and CE-QUAL-W2 models are programmed to calculate reaeration based on wind for lakes and on hydraulics for streams (and estuaries in the case of CE-QUAL-W2). The WASP and EPDRiv1 models calculate reaeration based on wind and hydraulics and have the option of summing the two or using the larger of the two calculated values.

5.3 Oxygen Demand

Oxygen demand generally refers to carbonaceous biochemical oxygen demand (CBOD) because any nitrogenous oxygen demand is modeled separately as ammonia nitrogen. Some models include the potential for modeling more than one CBOD, with separate first-order decay rates, considering that the CBOD may be more highly reactive from some sources (e.g., CSOs) and less reactive from other sources (e.g., groundwater inflow). In general, the more complex featured land-based water quality models can account for CBOD. The HSPF, SWAT, WARMF, and SWMM models will model CBOD in the same way described for simulation of nutrients and other pollutants, for both surface runoff and groundwater.

Except for BATHTUB, all the featured receiving water models represent oxygen demand. However, there are some differences in how the oxygen demand is simulated. Featured models, including QUAL2K, EPDRiv1, WASP, and HSPF, model the oxygen demand as CBOD and are subject to first-order decay and settling (SWMM is limited to first-order decay only). The EFDC, CE-QUAL-ICM, and WARMF model oxygen demand as organic matter. The CE-QUAL-W2 model is capable of modeling both CBOD and organic matter, so care must be taken not to "double dip" (i.e., represent the oxygen demand loads as both CBOD and organic matter, rather than splitting it between the two) if both are being modeled. Both CE-QUAL-ICM and EFDC model organic carbon to represent CBOD and do not have a state variable. These models also include a COD state variable that can account for processes such as oxidation of sulfide to sulfate.

5.4 Sediment

Most of the featured land-based models simulate solids in surface runoff (exceptions are VRRM and BMPTRAINS, which are simple models that focus on total nitrogen and total phosphorus). Many of the low to moderate complexity models apply a version of the universal soil loss equation (USLE) for sediment loads from pervious lands. The original USLE was formulated to determine annual soil loss based on a rainfall erosivity factor, soil erosivity factor, length and slope factor, cover management factor, and supporting-practice factor (e.g., contouring, terracing); a sediment delivery ratio was used to evaluate the amount of soil loss delivered to the receiving water. The modified USLE, applied in models such as SWAT, calculates daily sediment delivery based on daily runoff and peak daily flowrate. The revised USLE (RUSLE), applied in models such as Agricultural Non-Point Source Pollution Modeling System (AGNPS), similarly calculates daily soil loss based on daily runoff and peak daily flowrate, and then calculates a delivery ratio of eroded soil to the receiving water. More complex models including HSPF and WARMF can explicitly account for sediment detachment during storm events and reattachment during dry periods, with potential washoff of detached sediment during runoff events. Sediment loads for impervious areas typically apply an EMC to runoff or use buildup and washoff calculations to determine available sediment mass on the surface and washoff as a function of runoff intensity.

Receiving water processes may include streambank erosion, bed scour, and suspended solids settling. The featured models range from no process simulation to explicit simulation of cohesive (silt, clay) and non-cohesive (sand) particles. The EFDC, HSPF, and WARMF models all evaluate modeled parameters such as shear stress and velocity against a defined critical value to determine when settling, erosion, and scour are occurring. The Water Quality Analysis Simulation Program (WASP) can model multiple inorganic suspended solids, but it does not have a scour or erosion function that is dependent upon modeled velocity or shear stress. The CE-QUAL models all include a single inorganic suspended solid with a settling rate and no simulation of erosion or scour.

5.5 Nutrients

All the featured land-based models simulate nitrogen and phosphorus in surface runoff. The VRRM and BMPTRAINS models use total nitrogen and total phosphorus EMCs related to land-cover type. In most of the models, urban runoff can be modeled by either EMCs, buildup/washoff calculations, or both. Some of the more detailed models, particularly those that are best suited for agricultural evaluations, explicitly account for surface nitrogen and phosphorus mass within a modeled "surface layer" and determine the nutrient mass associated with surface runoff based on the concentrations of the sediment-sorbed and dissolved nutrients in that surface layer. The Generalized Watershed Loading Functions (GWLF) model offers a special routine that accounts for higher nutrient loads as a function of study area features such as tile drainage, septic tanks, manure application to agricultural fields, and animal grazing.

Groundwater loads of nutrients can also be an important factor, particularly in agricultural areas where elevated nitrate/nitrogen concentrations may be discharged to receiving waters. Again, the featured models that are best suited for agricultural evaluations explicitly account for subsurface concentrations of nutrients and can deliver the groundwater nutrient loads to the receiving water. The SWMM and GLWF models assign constant concentrations for groundwater nitrogen and phosphorus. The HSPF model offers the alternative of assigning monthly groundwater concentrations, which can be used to account for cyclical practices such as fertilization, if nutrients are not explicitly modeled.

As displayed in Table 5.8, the methods of addressing benthic nutrient fluxes in the featured receiving water models range from no simulated fluxes (QUAL2K, SWMM, WARMF, BATHTUB), to defined nutrient flux rates (EPDRiv1 and CE-QUAL-W2, HSPF), to a full simulation of sediment diagenesis (EFDC, WASP, CE-QUAL-ICM). The sediment diagenesis simulation accounts for the degradation of organic matter that settles to the bottom sediments, creating dissolved nutrients that accumulate in the sediment pore water and move to the water column through pore water exchange. The WASP model can use either sediment diagenesis calculations or assigned flux rates. The HSPF model allows for high and low release rates that depend upon the dissolved oxygen concentration in the receiving water (high rate used when dissolved oxygen drops below the assigned threshold dissolved oxygen value).

5.6 Bacteria

Bacteria are primarily a public health concern and may have a variety of both animal and human sources. Water quality standards may apply for one or more type including fecal coliform, total coliform, *Escherichia coli*, and *enterococci*. Surface runoff levels of bacteria tend to vary significantly on a seasonal basis, so a modeling approach that allows for seasonal runoff concentrations may be preferable. Standards often include a geometric mean standard as well as an instantaneous maximum, so a continuous simulation model would be required to truly assess the compliance with the standards.

Of the featured land-based models, those that can evaluate bacteria include GWLF, HSPF, AGNPS, SWAT, WARMF, and SWMM. Evaluation in the SWMM allows for modeling surface runoff bacteria by buildup–washoff or EMC by land-use category, but it does not allow for seasonal variability. The other models account for bacteria in essentially the same manner as nutrients. Again, GLWF offers a special routine that accounts for higher bacteria loads as a function of study area features such as septic tanks, manure application to agricultural fields, and animal grazing.

Receiving water processes are typically limited to first-order loss, which may include processes such as base die-off, light mortality, and settling. The SWMM first-order loss rate is constant, whereas others are typically adjusted for higher values as water temperature increases.

5.7 Phytoplankton, Periphyton, Macrophytes, and Zooplankton

Phytoplankton, periphyton, macrophytes, and zooplankton are receiving water constituents that are affected by factors such as availability of nutrients and light. Phytoplankton (typically measured based on chlorophyll-a content) are free-floating algae in receiving waters. Excessive phytoplankton in receiving waters may be called "algae blooms". In contrast, periphyton are algae attached to the bed sediment (typically on rocky bottoms and where there is sufficient light penetration). Macrophytes are attached plants, primarily on the fringes of a receiving water, that are rooted underwater, but extend above the water surface. Zooplankton are organisms that float in the water and feed on the phytoplankton.

The featured receiving water models offer different capabilities for phytoplankton, periphyton, macrophytes, and zooplankton. The SWMM does not model these constituents, and BATHTUB predicts chlorophyll-a as an indicator of phytoplankton. The other models all simulate phytoplankton and either periphyton or macrophytes (or both). The HSPF and CE-QUAL-W2 models also can explicitly model zooplankton. Regarding phytoplankton, QUAL2K, EPDRiv1, and HSPF are limited to a single phytoplankton, whereas the others generally offer three different phytoplankton, as discussed below.

The consideration of three phytoplankton species is designed to include cyanobacteria, diatom algae, and green algae. As indicated in Tetra Tech, Inc. (2007c), these algae will tend to exhibit the following different characteristics:

- Cyanobacteria, commonly called "blue-green algae", may use atmospheric nitrogen as a nutrient source.
- Diatoms are large algae with relatively high settling velocities and require silica as a nutrient to form cell walls; settling of spring diatom blooms may contribute to SOD in the summer.
- Green algae are generally those not falling into one of the other two categories, with the settling rate somewhere between the blue-green algae and diatoms.

Consequently, the simulation of the algae and the associated water quality constituents may benefit from modeling more than one species.

5.8 Metals

Metals can be toxic to aquatic life and may bioaccumulate in fish; they may be an indicator of pollution sources that may also be introducing other toxics. Most of the featured land-based models do not explicitly simulate any metals. Only WARMF and P8-UCM have one or more metals as a specific water quality constituent. The SWMM and HSPF models can account for runoff metals by simulating them as a "general pollutant"; to do so, the former model can use EMCs or buildup/washoff, while the latter can use buildup/washoff, potency factors relating constituent load to sediment load, or a combination of the two. Groundwater metals concentrations are specified as a constant value in SWMM and HSPF, while the HSPF has the option of setting monthly values. The WARMF is the most detailed model as it explicitly simulates metals in surface water and groundwater based on the modeled metals in the surface layer and subsurface.

Receiving water processes for metals may include attachment to sediment and subsequent settling with the sediment. The featured receiving water models, in general, have limited explicit modeling of specific metals. The WARMF model explicitly models a number of different metals. The various CE-QUAL models and EFDC explicitly simulate iron and manganese because those affect the adsorption and settling of dissolved nutrients. The WASP, EPDRiv1, CE-QUAL-W2, and HSPF models can model metals as a general constituent that can be simulated with processes including first-order decay and absorption to sediment and subsequent settling. The SWMM model can model metals as a general pollutant with first-order decay. The WARMF model also has the capability of simulating bioaccumulation of mercury in various fish species, and WASP has the capability of modeling metals in both the water column and bed sediment.

6.0 MODELING OF MANAGEMENT MEASURES

Models that focus on agricultural applications can evaluate a variety of practices including fertilizer and pesticide application rates and schedules, animal grazing and tilling practices, and crop harvesting. Models focusing on urban areas focus primarily on the benefits of structural BMPs that capture and treat surface runoff. The following provides summary information on the evaluation of management measures in water quality simulation for featured land-based models.

- AGNPS—Devices include field ponds; irrigation; drainage; tillage; fertilizer and pesticide application; and animal waste nutrient production.
- BMPTRAINS—Devices include retention basins, green roofs, vegetated natural buffers, stormwater harvesting, exfiltration trenches, tree wells, permeable pavement, swales, bioretention, filtering practices,

wet ponds, and user-defined BMP. Removal is based on device size relative to tributary area and associated runoff.

- GLEAMS—Devices include fertilizer and pesticide application, irrigation scheduling, and tillage operations.
- GWLF—Devices include assigned load reduction factors for detention ponds; infiltration basins; filter strips; stabilized stream banks and wetlands; animal waste management systems; street sweeping.
- HSPF—Devices include application of fertilizer and pesticides; BMP Module for simple percent load reduction; and disconnected impervious area. Guidance in HSPF Toolkit can be used to explicitly account for runoff flow control.
- OPUS—Devices include pesticide application, fertilization, tillage, and drainage (tile).
- P8-UCM—Devices include explicit modeling of detention ponds, infiltration basins, and swales; street sweeping is also included.
- SWAT—Devices include harvesting; grazing; tillage; fertilizer and pesticide application; and start/end of growing season. Street sweeping is included for urban areas.
- SWMM—Devices include assigned BMP removal efficiency by land-use category and the explicit modeling of low impact development (LID) features including bioretention, rain gardens, green roofs, infiltration trenches, permeable pavement, rain barrels, and swales.
- VRRM—Devices include vegetated roofs, rooftop disconnection, permeable pavement, grass channels, dry swales, bioretention, infiltration, extended detention ponds, sheetflow to filter/open space, wet swales, filtering practices, constructed wetland, wet ponds, and manufactured treatment devices.
- WARMF—Devices include buffer zones livestock fencing, detention ponds, and street sweeping.

The modeling of structural BMPs can range from simple load reduction percentages to explicit modeling of particular practices that will reflect the benefit based on the design of the facility. The BMPTRAINS model provides removal efficiencies for total nitrogen and total phosphorus based on the selected structural facility (or facilities) and the design of each facility (e.g., inches of treatment storage over the effective tributary impervious area). The VRRM model calculates annual load reductions of nitrogen and phosphorus based on specific design criteria established within the VRRM spreadsheet. In both cases, the average annual efficiency is based on previous long-term simulation evaluations that form the scientific basis for the

TABLE 5.10 Land-based water quality simulation model technical support and documentation.

Land-Based Model	Model Development Tools	Technical Support	Documentation
AGNPS	Data preparation and output tools	Unknown	Bingner, Theurer, & Yuan, 2018
BMPTRAINS	GUI data input and output	Workshops and online support page	Wanielista, Hardin, Kuzlo, & Gogo-Abite, 2014
GLEAMS	None	None	Knisel & Davis, 2000
GWLF	BASINS, MapShed	Unknown	Haith, Mandel, & Wu, 1992; Evans & Corrandini, 2011
HSPF	BASINS	Online user group (BASINS)	Bicknell, Imhoff, Kittle, & Donigian, 1996
OPUS	GUI (for Windows 95)	Unknown	USDA, 2000; USDA, 1992
P8-UCM	GUI for model input, output	Online help file	Walker, 1990
SWAT	GIS, data pre- and post-processing tools	Online user group (BASINS); workshops; U.S. EPA webinar	Neitsch, Arnold, Kiniry, & Williams, 2011; Arnold et al., 2012
SWMM	GUI for model input, output; proprietary GIS	Online user group	Rossman, 2015
VRRM	Spreadsheet-based input and output	Instructions in spreadsheet	VDEQ, 2016
WARMF	GIS interface	Online help system	Chen, Herr, & Weintraub, 2001

overall efficiencies. The P8-UCM model explicitly models several device types, and the SWMM has explicit simulation capabilities for several LID features such as bioretention, pervious pavement, and others. In the SWMM, the simulated percentage of surface runoff load reduction is generally proportional to the percentage of runoff captured and infiltrated to underlying native soil, although infiltrated water can be modeled explicitly as part of the SWMM groundwater routine and the potential increase in groundwater load can be evaluated. The SWMM, GWLF, and

TABLE 5.11 Receiving water model technical support and documentation.

Receiving Water Model	Model Development Tools	Technical Support	Documentation
BATHTUB	GUI for model input, output	No	Online: http://www.wwwalker.net/bathtub/help/bathtubWebMain.html
CE-QUAL-ICM	Unknown	Unknown	Cerco & Cole, 1995
CE-QUAL-W2	GUI for data input and output	Online forum	Cole & Wells, 2015
EFDC	Proprietary pre- and post-processor	U.S. EPA webinars, workshops	Tetra-Tech, Inc., 2007a, 2007b, 2007c
EPDRiv1	GUI for model input, output (EPDRiv1)	Unknown	Environmental Laboratory, 1990; Martin & Wool, 2002
HSPF	BASINS	Online user group (BASINS)	Bicknell et al., 1996 Duda et al., 2012
QUAL2K	Spreadsheet-based input and output	Online user group	Chapra, Pelletier, & Tao, 2008
SWMM	GUI for model input, output; proprietary GIS	Online user group	Rossman, 2015
WARMF	GIS interface	On-line help system	Chen et al., 2001
WASP	GUI for data input and output	U.S. EPA webinars, workshops	Ambrose, Knightes, & Wool, n.d.; Wool, Ambrose, & Martin, n.d.; Wool, Ambrose, & Martin, n.d.; Wool, Ambrose, Martin, & Comer, n.d.

HSPF models all offer the option of a simple percent load reduction for areas served by BMP facilities.

Explicit modeling of surface runoff controls may provide more insight into surface runoff and base flow receiving water impacts regarding issues such as water temperature and streambank erosion and scour. Surface runoff from impervious land areas such as streets, parking lots, and roofs are likely to be warmer than surface runoff from undeveloped areas, and may result in localized high receiving water temperatures that adversely affect aquatic life. Typical peak runoff flow controls for development are expected to limit the occurrences of receiving water flows and velocities that may result in streambank erosion and scour.

7.0 OTHER MODEL SELECTION CONSIDERATIONS

Tables 5.10 and 5.11 summarize the evaluation of the featured land-based and receiving water models, respectively, for a few additional model features that may affect the selection of one or more models. These factors include the availability of model documentation and user support, tools for model development (e.g., data pre-processors and post-processors), and model support.

8.0 REFERENCES

Ambrose, R. B., Knightes, C., & Wool, T. A. (n.d.). *WASP7 pH-alkalinity—model theory and user's guide, supplement to WASP user documentation.* Atlanta, GA: U.S. Environmental Protection Agency.

Arnold, J. G., Kiniry, J. R., Srinivasan, R., Williams, J. R., Haney, E. B., & Neitsch, S. L. (2012). *Soil and water assessment tool input/output documentation version 2012*, (Texas Water Resources Institute Technical Report No. 439). College Station, TX: Texas Water Resources Institute.

Bicknell, B. R., Imhoff, J. C., Kittle, J. L., & Donigian, A. S. (1996). *Hydrological simulation program—FORTRAN user's manual for release 11.* Athens, GA: U.S. Environmental Protection Agency Environmental Research Laboratory.

Bingner, R. L., Theurer, F. D., & Yuan, Y. (2018). *AnnAGNPS technical processes documentation version 5.5.* Oxford, MS: United States Department of Agriculture-Agricultural Research Service.

Burton, G. A., & Pitt, R. E. (2001). Appendix H, watershed and receiving water modeling. In *Stormwater effects handbook: A toolbox for watershed managers, scientists, and engineers.* Boca Raton, FL: CRC Press.

Cerco, C. F., & Cole, T. (1995). *User's guide to the CE-QUAL-ICM three-dimensional eutrophication model release version 1.0* (U.S. Army Corps of Engineers Technical Report EL-95-15). Vicksburg, MS: U.S. Army Engineer Waterways Experiment Station.

Chapra, S., Pelletier, G., & Tao, H. (2008). *QUAL2K: A Modeling framework for simulating river and stream water quality: Documentation and user's manual* (version 2.11). Civil and Environmental Engineering Department, Tufts University, Medford, MA.

Chen, C., Herr, J., & Weintraub, L. (2001). *Watershed Analysis Risk Management Framework (WARMF): Update one—A decision support system for watershed analysis and total maximum daily load calculation, allocation and implementation.* Palo Alto, CA: Electric Power Research Institute.

Cole, T. M., & Wells, S. A. (2015). *CE-QUAL-W2: A two-dimensional, laterally averaged, hydrodynamic and water quality model, version 3.72 user manual.* Portland, OR: Portland State University.

Donigian, A. S., Bicknell, B. R., & Love, J. T. (n.d.). *Watershed and receiving water model linkage with HSPF: Issues and example applications.* Retrieved from http://www.aquaterra.com/resources/pubs/pdf/donigian-2007b.pdf

Duda, P. B., Hummel, P. R., Donigian, A. S., & Imhoff, J. C. (2012). BASINS/HSPF: Model use, calibration, and validation. *Transactions of the ASABE, 55*(4).

Environmental Laboratory. (1990). *CE-QUAL-RIV1: A dynamic, one-dimensional (longitudinal) water quality model for streams: User's manual* (Instruction Report E-90-1). Vicksburg, MS: U.S. Army Engineer Waterways Experiment Station.

Evans, B. M., & Corradini, K. J. (2011). *Guide to utilizing the GWLF-E plug-in within the BASINS 4.0 environment.* University Park, PA: Penn State University.

Haith, D. A., Mandel, R., & Wu, R. S. (1992). *GWLF generalized watershed loading functions version 2.0 user's manual.* Ithaca, NY: Cornell University.

Knisel, W., & Davis, F. (2000). *GLEAMS: Groundwater loading effects of agricultural management systems, version 3.0 user manual.*

Martin, J. L., Ambrose, R. B., & Wool, T. A. (n.d.). *WASP7 Benthic algae—model theory and user's guide, supplement to WASP user documentation.* Atlanta, GA: U.S. Environmental Protection Agency.

Martin, J. L., & Wool, T. (2002). *A dynamic one-dimensional model of hydrodynamics and water quality EPD-RIV1 version 1.0 user's manual.* Atlanta, GA: Georgia Environmental Protection Division.

Martin, J. L., & Wool, T. A. (2012). *WASP7 sediment diagenesis routines: model theory and user's guide.* Atlanta, GA: U.S. Environmental Protection Agency.

Neitsch, S. I., Arnold, J. G., Kiniry, J. R., & Williams, J. R. (2011). *Soil and water assessment tool theoretical documentation version 2009*. Texas Water Resources Institute (Technical Report No. 406). College Station, TX: Texas Water Resources Institute.

Rossman, L. A. (2015). *Storm water management model user's manual version 5.1* (EPA-600/R-14/413b). Cincinnati, OH: U.S. Environmental Protection Agency.

Tetra Tech Inc. (2007a). *The environmental fluid dynamics code theory and computation volume 1: Hydrodynamics and mass transport*. Fairfax, VA: Tetra Tech Inc.

Tetra Tech Inc. (2007b). *The environmental fluid dynamics code theory and computation volume 2: Sediment and contaminant transport and fate*. Fairfax, VA: Tetra Tech Inc.

Tetra Tech Inc. (2007c). *The environmental fluid dynamics code theory and computation volume 3: Water quality model*. Fairfax, VA: Tetra Tech Inc.

U.S. Environmental Protection Agency. (2001). *PLOAD version 3.0: An arcview GIS tool to calculate nonpoint sources of pollution in watershed and stormwater projects user's manual*. Washington, D.C.: U.S. Environmental Protection Agency.

Virginia Department of Environmental Quality. (2016). *Guidance memo no. 16-2001—Updated Virginia runoff reduction method compliance spreadsheets—Version 3.0*. Richmond, VA: Water Permitting Division, Virginia Department of Environmental Quality.

Walker, W. W. (1990). *P8 urban catchment model program documentation* (version 1.1). Providence, RI: Narragansett Bay Project.

Wanielista, M., Hardin, M., Kuzlo, P., & Gogo-Abite, I. (2014). *User's manual for the BMPTRAINS model*. Orlando, FL: Stormwater Management Academy.

Wool, T. A., Ambrose, R. B., & Martin, J. L. (n.d.). *WASP7 multiple algae—Model theory and user's guide, supplement to WASP user documentation*. Atlanta, GA: U.S. Environmental Protection Agency.

Wool, T. A., Ambrose, R. B., & Martin, J. L. (n.d.). *WASP7 temperature and fecal coliform—Model theory and user's guide, supplement to WASP user documentation*. Atlanta, GA: U.S. Environmental Protection Agency.

Wool, T. A., Ambrose, R. B., Martin, J. L., & Comer, E. A. (n.d.). *Water quality analysis simulation program (WASP) version 6.0 draft user's manual*. Atlanta, GA: U.S. Environmental Protection Agency.

6

Models of the Future

Henry Manguerra, Ph.D., P.E. and Ayman Alafifi, Ph.D.

1.0 BACKGROUND

Watershed, water quality, and stormwater models have naturally evolved over the years to address emerging scientific, research, and business needs of the modeling community. However, the significant advances during the past two to three decades were mostly in the form of enhancement of the usability of existing models—understandably as the result of rapid advances in computing technology. Such technological advances have also resulted in the proliferation of integrated modeling systems that feature a suite of existing models with functional capabilities that complement one another to allow for more seamless and cost-efficient modeling and analysis of larger and more complex integrated water systems (e.g., interconnected watersheds, multiple receiving waters, and interactions with groundwater). Yet, as science tends to mature more deliberately, and fewer new scientific discoveries are being made relating to water models, the models of the future will be defined by advances in technology anticipated to far outpace advances in science. It is possible, though, that with adequate resources and commitment of the modeling community, fundamental developments in models to include scientific advances could arrive faster to address the more immediate and emerging business needs. Some of these business drivers, as identified in Chapter 3, include new stormwater management regulations, emerging contaminants and pathogens, the need for holistic and integrated management of water (i.e., One Water approach), extreme weather events, and the changing nature of pollution problems.

This chapter discusses the future of model development (i.e., models of the future), namely following two development trajectories:

- Anticipated "normal" advancement trajectory as a result of current and anticipated advances in technology and data, and
- Desired trajectory as driven by current and anticipated business needs.

2.0 MODEL DEVELOPMENT TRENDS AND OPPORTUNITIES

Models will improve as a direct result of advances in technology and data. These improvements will address the constant business need for models that are more user friendly, computationally efficient, comprehensive, and applicable to a wide range of applications.

2.1 Model Performance and Usability

Rapid advances in technology have improved model speed and performance, model usability, and, consequently, the models' ability to handle complex model applications. Nonetheless, many existing models remain difficult to use and require steep learning curves and a high level of expertise and effort for setup, preparation of input data, and analysis of simulation results.

Although model developers have invested time and resources in improving their model graphical user interfaces (GUIs), many GUIs remained fundamentally old-fashioned and characterized by traditionally form-based

GUIs with mapping and visualization integrated to varying extents. These GUIs will have to evolve creatively, from traditionally rigid forms to user-friendly and smart platforms with which users, regardless of their skill levels, can easily and naturally interact.

Thus, future GUIs will become more intuitive to a broader range of users, ranging from the experienced modelers to the novice; more easily modularized depending on the complexity and size of the system being modeled; and better able to import and process data seamlessly from external data sources when needed. The "TurboTax analogy" has been mentioned in the past in which users can either use the traditional form-based methodology or answer a series of questions to build and run their models.

There is also the need for models to adhere to a consistent look and feel. This could be achieved through a standard library of GUI components that can be reused to build the GUI of new models and rebuild that of the legacy models. This would include Web-based models that take advantage of advances in Web technology, providing more user-friendly click, drag, and drop environments to run models and visualize results. Having a library of standardized GUI components that can be reused and configured will also benefit models that do not have adequate time and resources to improve their GUIs.

2.2 Input Data Pre-Processing and Visualization of Results

Advances in models include the ability to process large sets of spatial and temporal data, including continuous real-time monitoring data that are becoming more prevalent as a result of advances in sensor technology. Although a considerable amount of real-time water data are available on the Web, their use for modeling is not fully exploited as modelers often have to spend significant effort to download and pre-process the data before they can be used. Some models have addressed this challenge by developing functional features to automatically download the data, check for data quality issues, and format them for model input. With more data becoming accessible via the Internet, it is anticipated that these functionalities will be a common feature of the models of the future. Laniak et al. (2013) emphasized a similar need for tools and methods for (a) automated discovery, access, and retrieval of data from individual sources and (b) ensuring quality and facilitating integration of data from multiple sources and populating model input files. Advances in Internet programming (e.g., use of open source programming languages such as R and Python) and data analytics software could provide new techniques that can handle a wide range of disparate data sources.

The ability of water resources models to provide easy-to-interpret and actionable results that can be communicated to decision-makers and the public largely depends on post-processing and visualization of model results. Traditionally, this task falls into the hands of modelers to subjectively filter through large sets of model outputs to generate meaningful tables, graphs, and maps that can be used to inform decision-makers. Communicating model results would be improved by providing model users the ability to run the models on their own and to interactively adjust input parameters and visualize the outputs. Several tools have been developed to assist model developers in building such interactive, dynamic, multidimensional plots such as R Shiny package and Python Plotly. In addition, the use of geographic information system (GIS) online platforms, such as Esri ArcGIS Online, allow model developers to publish and visualize large sets of spatial and temporal data and provide model users access to model data and outputs.

2.3 Integration With Data Collection and Management Systems

The integration of GIS with models to pre-process and prepare spatial data for model input was a significant advancement over the last two decades. A number of new models have also used relational databases (including geodatabases) in lieu of or in combination with traditional individual files to store both input and output model data. Although models of the future will continue to take advantage of incremental advances in GIS and data management technologies, the more significant advancement in models will focus on how to effectively take advantage of the proliferation of large amounts of Internet-connected spatial and temporal data. For example, large amounts of remotely sensed data such as light detection and ranging (LiDAR) and multispectral imagery are being collected and are made accessible for consumption. Advances in sensor technology have also spurred the collection of massive amounts of continuous monitoring data. These datasets, which are often referred to as "big data", are characterized as datasets that are too large or complex (in terms of volume, velocity, and variety) for analysis using traditional analytical and numerical techniques. The challenge for the modeler is how to take advantage of big data within the current model implementation. Advances in big data analytics open up the possibility for enhancing the capability of existing models and the development of new data-driven models with descriptive, diagnostic, predictive, and prescriptive applications for watershed, water quality, and stormwater management (Paul, Manguerra, & Slawecki, 2018).

2.3.1 *Integrated Modeling Systems*

Recent decades have seen a proliferation of models designed for different purposes simulating various hydrologic-hydraulic-water quality processes at different spatial and temporal scales, in addition to program-specific integrated modeling systems that attempted to make these models "talk" to one another. This rich, diverse, but disorganized modeling landscape has led Laniak et al. (2013) to suggest the importance of a common framework for developing, testing, and maintaining a suite of community-level models or model components. Moreover, because modelers often find it necessary to use and integrate multiple models for more complex water systems, the need for a common framework becomes even more compelling because models following a common framework are anticipated to be more inherently integrable. The ability to better manage the exploding amount of information has been a recurring need for many years now and has resulted in the emergence of various technology solutions (including the latest emerging technology of Enterprise Taxonomy and Ontology Management) to manage content, document, metadata, and knowledge. It is anticipated that similar technology solutions will become available to support the need for a common modeling framework.

2.3.2 Model Documentation and Accessibility

The open source movement that began several decades ago not only made the source code and documentation of open source models freely available, but also added additional pressure for both proprietary and public-domain "private" models for transparency and increased accessibility. The relatively recent emergence of online model development framework and repositories for both open source and proprietary software such as SourceForge, GitHub, BitBucket, and HydroShare is also helping address the ever-increasing need for models to be accessible and well documented. This open source trend is expected to continue and become even more widespread, adding more public domain models to become open source. The need for data and code transparency is also slowly finding its way into publication policies and guidelines for a number of water resources peer-reviewed publications. For example, the *Journal of Environmental Modeling and Software* highly encourages authors to submit their data or upload them into a shareable URL or online repository. The American Society of Civil Engineers *Journal of Water Resources Planning and Management* in September 2018 issued a new policy to specify available data, models, and code for submitted manuscripts. Their open-access journal *Data in Brief* was launched to provide authors a way to share and reuse their datasets to facilitate reproducibility of their findings and encourage collaboration.

3.0 MODEL DEVELOPMENT GAPS AND NEEDS

While advances in technology will naturally improve models' usability, computational speed, and performance, fundamental model developments will become increasingly deliberate and will be driven by known and anticipated functional/discipline gaps and business needs. Functional/discipline gaps, as discussed in this chapter, range from the need to improve the scientific rigor of models and model components to address known functional capability gaps. Business needs include the need for more comprehensive and holistic models that incorporate the interrelationships and dynamics, not only among the various subsectors and disciplines within water resources, but also with other sectors such as agriculture, energy, and the environment; the need for practically useful models for planning, decision support, real-time operation and maintenance, and disaster forecasting, preparation, and mitigation; and the need for models that allow for and take advantage of distributed and participatory modeling.

3.1 Functional/Discipline Gaps

Scientific rigor and functional capabilities of existing models are expected to improve through (a) enhancement of existing model components, (b) development of new model components, and (c) integration with other models. Entirely new models are also expected to emerge, although more likely as either open source or proprietary versions of the original public-domain models and/or as the result of the integration of models.

3.1.1 Scientific Rigor

The improvement in scientific rigor, especially of mechanistic models, is a natural result of advances in scientific research (Daniel, Camp, LeBoeuf, Penrod, Dobbins, & Abkowitz, 2011) and computational techniques. Mechanistic models are built by modeling individual yet coupled components and processes of the entire water system. These include hydrologic, hydraulic, sediment, water quality, and ecological processes over a range of spatial and temporal domains. A number of these processes, such as hydrologic and hydraulic processes, are relatively well understood, and any scientific improvements of these processes will be small compared to improvements expected or needed for the other processes. Therefore, it is anticipated that more scientifically rigorous model components will be produced in the future for such processes including erosion, sediment transport, and deposition; fate and transport of currently less understood water quality constituents; and ecological response to hydrologic, hydraulic, and water quality changes. Advances in computational techniques and computing power will also allow systems to be modeled in two or three dimensions. This direction is already clearly evident by recent model improvements, as demonstrated by the following examples:

- The U.S. Department of Defense and the U.S. Army Corps of Engineers (USACE) have developed sediment transport models for Great

Lakes tributaries and have expressed their commitment to continue these efforts in the future (USACE, 2008).

- The USACE recently completed and is continuing to enhance the sediment transport and water quality module for the HEC-RAS model (USACE 2016, 2018).
- The Soil and Water Assessment Tool (SWAT) and Hydrologic Simulation Program FORTRAN (HSPF) have provided users with the ability to integrate remote sensing data to more accurately characterize runoff, determine flow paths, and quantify water quality constituents downstream of a point or a non-point source (Gao & Li, 2014).
- The United States Geological Survey (USGS) developed the Physical Habitat Simulation System (PHABSIM) to help manage flow to improve habitat. The PHABSIM simulates the relationship between river flood components (e.g., magnitude, frequency, duration, etc.) and physical habitat for various life stages of indicator and key riverine species.

It is also expected that improvement in the scientific rigor of models is also the result of improvements in model calibration and validation, parameter estimation, uncertainty analysis, and numerical computational techniques. Examples include the following:

- Coupling deterministic models with statistical tools such as Monte Carlo (Van der Lee, Van der Molen, Van den Boogaard, & Van der Klis, 2006), bootstrap sampling (Ayllon, Almodovar, Nicola, & Elvira, 2012), Bayesian networks and fuzzy numbers (Burgman, Breininger, Duncan, & Ferson, 2001) to perform uncertainty analysis to better capture system variability.
- Using data mining techniques such as supervised cluster analysis (Chen et al., 2017) and data assimilation methods such genetic evolutionary algorithms (Shin, Mohanty, & Ines, 2018) to save numerical computational costs to explore historic patterns and forecast future trends.
- Improving and automating parameter estimation capabilities to compare outputs to historical observations. Examples of these improvements are evident in the USACE Hydrologic Engineering Center's Hydrologic Modeling System (HEC-HMS) (SKahill, 2006) and Water Evaluation And Planning System (WEAP) parameter estimation tools.
- Moving beyond 1-dimensional models to incorporate two and three dimensions. For example, USACE HEC River Analysis System

HEC-RAS (Version 5.0) recently included two-dimensional capabilities to better capture flow patterns and flood impacts, particularly for urbanized areas, levee failure analysis, and wetlands ecological studies.

New models or model applications will also be the result of integration of models (i.e., coupling a watershed model with a receiving water model or a hydrologic/hydraulic model with a water quality model). The need to maintain and validate scientific rigor when integrating models has been emphasized during the International Summit on Integrated Environmental Modeling (U.S. EPA, 2013). This can be achieved by (a) reducing occurrence and propagation of errors when models are linked and (b) reconciling issues related to different scales and temporal/spatial resolutions and uncertainties of the linked models.

3.1.2 Functional Gap—Ecological Response

Simulating ecological response to hydrologic, hydraulic, and water quality changes in the water system remains a major functional gap. Current approaches have been primarily data driven (i.e., supplemented by subject matter expert knowledge and opinions) to develop stressor-response relationships between paired indicators through empirical, statistical, and heuristic means. Hence, the use of ecosystem models for decision-making remains very limited and, therefore, future advancements should include the capability to simulate biological interactions to evaluate water quality (Goethals & Forio, 2018; Horn, Rueda, Hormann, & Fohrer, 2004). It is anticipated that future ecosystem models will be characterized by the following:

- More seamless integration with hydrologic, hydraulic, and water quality models;
- Better methods to identify key indicator species and quantify their ecological response to changes in flow components;
- Better tools to mathematically quantify the intricate inter- and intra-species relationships and their water-related habitat needs;
- Mechanistic representation of impacts in relation to land use, urban development, infrastructure implementation, and ecosystem services; and
- Integration of optimization tools to better capture a wide range of management actions that can help improve habitat quality for key indicator species.

The combined physical–water-quality–ecosystem models are also anticipated to be complex to use and, therefore, user interfaces and tools that enhance their transparency, understanding, and user-friendliness are important. Transparency and clarity of model limitations with respect to the extent of model calibration/validation, uncertainty, and applicability are also expected to significantly improve.

3.1.3 Functional Gap—Ecosystem Services and Socio-Economic Impact Assessment

Although assessment of ecosystem services and, subsequently, socio-economic impacts have been conducted successfully in the past, they were completed with varying scientific rigor, scales, and resolutions. Advances in ecosystem models and their linkage with hydrologic, hydraulic, and/or water quality models will allow for more seamless and standardized extension of the modeling exercise toward the assessment of ecosystem services and socio-economic impacts. These models should be able to perform analyses such as those identified by the U.S. Environmental Protection Agency (U.S. EPA) (2018):

- How to estimate current production of ecosystem goods and services, given the type and condition of ecosystems,
- How ecosystem services contribute to human health and well-being, and
- How the production and benefits of these ecosystem services may be reduced or sustained under various decision scenarios and in response to regional conditions.

Once these basic functional needs are satisfactorily addressed by these models, then issues related to scales, resolutions, and applicability can be addressed. Linkage most likely will involve models of different scales and resolutions, and, therefore, the need to scale up, scale down, or adopt nested resolutions will have to be addressed. According to Kling et al. (2016), this presents a fundamental challenge for model developers to address the tradeoff between model tractability and the level of detail with which the economic behavior, crop growth, and watershed processes are modeled. According to the authors, innovations are needed to better incorporate dynamic aspects of economic behavior into integrated models that are "realistic enough" in their representation of component processes. They also noted that it is important to quantify and minimize the aggregation bias and efficiency loss caused by differing spatial scales at which economic and biophysical processes operate.

3.1.4 Functional Gap—Best Management Practices Integration

Best management practices (BMPs) have been integrated in models either as point sinks through the use of performance efficiencies or as explicit elements of a landscape/conveyance/transport system, where the fundamental processes of detention, infiltration, evaporation, and others are simulated to assess the BMP treatment performance. Regardless of the methodology used, future models are anticipated to be able to include a wider array of BMPs (including proprietary BMPs) that can be seamlessly integrated in the simulation. For models that are designed to simulate the fundamental BMP processes, the ability to easily define the BMP physical configuration and materials are also expected. Because BMPs are modeled to support BMP performance comparison and selection, these models are also expected to easily compare and contrast alternative BMP scenarios.

It is uncertain whether the future points to a direction where BMPs are handled mechanistically within the water system as opposed to being considered as black boxes or point sinks. There will always be BMP types that can be alternatively simulated mechanistically and others that can be as effectively and accurately modeled as point sinks. Separate advances in each of these methodologies are anticipated, such as those identified by Roesner (2013); anticipated advances include the ability to estimate uncertainties involved in using removal efficiencies and BMP performance algorithms that are practical yet represent the water quality behavior of BMPs. Recent efforts by U.S. EPA resulted in developing the Green Infrastructure Flexible Model (GIFMod) (Massoudieh et al., 2017), which can be used to build conceptual models of green infrastructures to predict their hydraulic and water quality performance. This model provides users more control over the parameters defining flow, transport of water quality constituents, reaction rates, and stoichiometry.

3.1.5 Functional Gap—Integration of Global Climate Change Models

Before watershed-specific models can use the climate change projections from the global general circulation models, these projections require adjustment using a statistical downscaling technique to address the scale differences between the large-scale climate features from the local climate of the area of interest. Currently, the selection of the global circulation model and the downscaling technique are done offline and primarily rely on the modeler's ability to produce the downscaled climate data correctly. There is a need to make this linkage to climate data more seamless, possibly by making the models geographically aware and subsequently capable of providing the user a menu of global circulation models and downscaling techniques that are appropriate for the geographic area of interest. Additional enhancements will include allowing the modeler to generate local climate data in a format that is readily consumable by the model without leaving the model framework.

3.2 Integration Within Water Resources and With Other Sectors

Watershed, stormwater, and water quality models are anticipated to be coupled with other models in an integrated modeling framework to address the need for comprehensive yet rigorous models that can be used for holistic water resources planning and management following integrated water resources management (IWRM) principles or the One Water approach. Roesner (2013) noted, for instance, that an integrated urban water system model that can handle water supply, wastewater, and stormwater and the linkages among them will be beneficial for identifying opportunities to reduce water use, utility costs, energy requirements, and wastewater flows, among other benefits. Manyanhaire (2014) further noted that such models should address the interconnectedness of people, ecosystems, hydrology, and many other factors at the basin context to adequately address complex issues related to environmental benefits, impacts on agro-economies, and water security.

These models are also anticipated to be part of broader modeling frameworks encompassing multiple sectors. Kling et al. (2016), for example, have articulated the importance of an integrated assessment model for the food, energy, and water (FEW) nexus. According to the authors, the integrated assessment model will involve the integration and use of crop models, economic models for land use, watershed/water quality models, and bioenergy models.

As future IWRM and FEW modeling frameworks will likely involve integrating existing models, the challenges noted earlier relating to propagation of errors and uncertainties when linking models of varying mathematical rigor, system boundaries inputs and outputs, scales, coverages, and resolutions will have to be addressed. Stormwater models of the future should consider these challenges to facilitate their integration and use within the modeling frameworks.

3.3 Planning Models/Decision Support Systems

Models are traditionally used for planning and decision-making by simulating and comparing the results of alternative model scenarios. However, there is often a strong disconnect between model results, which are often presented by the modelers closer to their raw and complex form, and the synthesized information that decision-makers can use as the basis for their planning and decision-making. It is anticipated that future models will feature a dashboard that will allow modelers, decision-makers, and stakeholders to mine, synthesize, and visualize both scenario data and the results to support meaningful planning and decision-making, if needed, in a multi-objective and multi-stakeholder fashion. According to Laniak et al. (2013), because stakeholders are diverse with different disciplinary backgrounds, vocabularies, priorities, and vested interests, various synthesis and visualizations may have to be made available to suit particular audiences and will have to be designed hierarchically to move seamlessly from very general displays of overall results to highly detailed, component-based visualizations.

3.4 Real-Time Operation, Maintenance, and Disaster Mitigation

With advances in sensor and information technology, an enormous amount of real-time monitoring data can be acquired, processed, interpreted, and used to operate and maintain water systems and forecast, prepare for, and mitigate disaster events such as flooding and water quality contamination. Future watershed, stormwater, and water quality models are anticipated to take advantage of the availability of this enormous amount of data. This includes satellite data that are becoming more readily available for end users' consumption. Advances in related technologies such as big data analytics, artificial intelligence, and the Internet of Things will also spur the emergence of data-driven models (as opposed to mechanistic models) to predict system responses based on past memories' cause-and-effect patterns. An example of this is the use of artificial neural networks for forecasting streamflows and rainfall-runoff on watersheds (Daniel et al., 2011).

3.5 Distributed and Participatory Modeling

Advances in cloud computing, data connectivity through the Internet, and Web-based user interfaces and visualization are paving the way for distributed and participatory modeling to be more feasible and practical. Distributed modeling can be thought of as similar in concept to distributed computing and crowd sourcing combined, where distributed expertise and computing resources are taken advantage of to complete a modeling task. A large modeling task can be broken down geographically and/or functionally into smaller jobs, with each job completed by the appropriate expert either incrementally or concurrently with other jobs. The experts collaborate, as needed, to complete the modeling task. Distributed modeling presents an opportunity to promote a more meaningful and dynamic participation from the stakeholders during the modeling exercise as opposed to a situation where the stakeholders are informed of the modeling results after the fact. This may lead to efficiencies both in terms of speed and cost of identifying and assessing solutions that are mutually beneficial and acceptable to the stakeholders.

4.0 SUMMARY AND RECOMMENDATIONS

Advances in data and technology are expected to result in models of the future that are more user friendly, computationally efficient, scientifically rigorous, and comprehensive in functional scope, allowing for their use in a wide range of applications. Such applications are expected to serve the needs of multidisciplinary teams and address the ever-constant need for the cost-effective and practical use of these models in all facets of planning, operation, maintenance, and decision-making. These models are also expected to be flexible, modular, and adaptable for simple to complex water systems. It is certain that the models will continue to evolve as business needs will invariably change as a result of new regulations and policies, new contaminants, scientific discoveries, shifts in business paradigms, technology breakthroughs, and improved techniques and approaches. As such, there is a need for the watershed, water quality, and stormwater modeling community to formally organize, share knowledge and resources, and collaborate toward achieving a model development trajectory that fully takes advantage of advances in data and technology, but also steers model development toward addressing the most pressing business needs. The following next steps are, therefore, presented for consideration:

1. Organize formally. This document could represent an important justification to formally organize, beginning with the basic purpose of updating and publishing its contents on a regular basis, for instance, or, at a minimum, through the Internet. This document could potentially become the "living encyclopedia" of watershed, water quality, and stormwater models and provide the modeling community a common knowledge and understanding of its current state. With this common knowledge and understanding, the modeling community could be persuaded more easily to work together toward steering and coordinating future model developments to achieve common business needs. The call to organize can be addressed initially through the formation of a small group that will serve as the "owners" of this document. In time, and when it is practically feasible, the group could evolve into a "Community of Practice" with broader collective responsibility to advance the state of modeling science and practice.
2. Establish an online one-stop knowledge shop. In addition to the document updates, the online one-stop knowledge shop can serve as a gateway to a resource of information about watershed, water quality, and stormwater models. For example, this knowledge website could provide links to additional detailed documentation of the models. Related documents on model selection, calibration and validation, model case studies, and other best practices can also be provided.

3. Plan, execute, and monitor to achieve the desired model development trajectory. It is important that the plan include specific short- and long-term outputs that are clearly attainable given the resources and the time. The objective is to advance the state of the science and the practice of watershed, water quality, and stormwater modeling to a development trajectory toward addressing the most pressing business needs while taking advantage of advances in data and technology. As funding and resources are needed, it is important to obtain institutional buy in and support of this activity. Such support could come from the Water Environment Federation, research institutions such as the Water Research Foundation, and/or federal agencies such as U.S. EPA, the United States Department of Agriculture, and the USGS. It is expected however, that the professional community in the public and the private sectors will continue to volunteer and contribute a significant portion of the effort required.

5.0 REFERENCES

Ayllon, D., Almodovar, A., Nicola, G. G., & Elvira, B. (2012). The influence of variable habitat suitability criteria on phabsim habitat index results. *River Research and Applications, 28*(8), 1179–1188.

Burgman, M. A., Breininger, D. R., Duncan, B. W., & Ferson, S. (2001). Setting reliability bounds on habitat suitability indices. *Ecological Applications, 11*(1), 70–78.

Chen, D., Leon, A. S., Hosseini, P., Gibson, N. L., & Fuentes, C. (2017). Application of cluster analysis for finding operational patterns of multi-reservoir system during transition period. *Journal of Water Resources Planning Management, 143*(8). doi:10.1061/(ASCE)WR.1943-5452.0000772

Daniel, E. B., Camp, J. V., LeBoeuf, E. J., Penrod, J. R., Dobbins, J. P., & Abkowitz, M. D. (2011). Watershed modeling and its applications: A state-of-the-art review. *The Open Hydrology Journal, 5*, 26–50

Gao, L., & Li, D. (2014). A review of hydrological/water-quality models. *Frontiers of Agricultural Science and Engineering, 1*(4), 267–276.

Goethals, P. L. M., & Forio, M. A. E. (2018). Advances in ecological water system modeling: Integration and leanification as a basis for application in environmental management. *Water, 10*(9), 1216.

Horn, A. L., Rueda F. J., Hormann, G., & Fohrer, N. (2004). Implementing river water quality modeling issues in mesoscale watershed models for water policy demands—An overview on current concepts, deficits, and future asks. *Physics and Chemistry of the Earth, 29*(11–12), 725–37.

Kling, C. L., Arritt, R. W., Calhoun, G., Keiser, D. A., Antle, J. M., Arnold, J., ... & Zhang, W. (2016). "Research needs and challenges in the FEW system: Coupling economic models with agronomic, hydrologic, and bioenergy models for sustainable food, energy, and water systems." White paper prepared for the National Science Foundation's Food, Water Workshop, Iowa State University, October 11–12, 2015.

Laniak, G. F., Olchin, G., Goodall, J., Voinov, A., Hill, M., Glynn, P., ... & Hughes, A. (2013). Integrated environmental modeling: A vision and roadmap for the future. *Environmental Modelling and Software, 39*, 3-23.

Manyanhaire, I. O. (2014). Interrogating the applicability of the IWRM model. *Environment and Natural Resources, 4*(1).

Massoudieh, A., Maghrebi, M., Kamrani, B., Nietch, C., Tryby, M., Aflaki, S., & Panguluri, S. (2017). A flexible modeling framework for hydraulic and water quality performance assessment of stormwater green infrastructure. *Environmental Modelling & Software*, *92*, 57–73.

Paul, S., Manguerra, H. B., & Slawecki, T. (2018). The state of the art of big data analytics: A watershed management perspective. *Proceedings of the Annual Water Environment Federation Technical Exhibition and Conference,* New Orleans, Louisiana, United States, September 29–October 3. Alexandria, VA: Water Environment Federation.

Roesner, L. A. (2013). A framework for watershed modeling & the need for an integrated urban water system model. In Field, R., & Baker, M., Jr. (Eds.), *Fifty years of watershed modeling: Past, present and future.* Proceedings of the ECI Symposium Series. Retrieved from https://dc.engconfintl.org/watershed/

Shin, Y., Mohanty, B. S., & Ines, A. (2018). Development of non-parametric evolutionary algorithm for predicting soil moisture dynamics. *Journal of Hydrology, 564*, 208–221.

Skahill, B. E. (2006). *Potential improvements for HEC-HMS automated parameter estimation* (ERDC/CHL TR-06-13). Vicksburg, MS: Coastal and Hydraulics Laboratory, U.S. Army Engineer Research and Development Center.

U.S. Army Corps of Engineers. (2008). Great Lakes tributary modeling program. Available from US Army Corps of Engineers (USACE), Great Lakes and Ohio River Division.

U.S. Army Corps of Engineers. (2016). *HEC-RAS river analysis system user's manual* (Version 5.0). Davis, CA: U.S. Army Corps of Engineers Hydrologic Engineering Center.

U.S. Army Corps of Engineers. (2018). *HEC-RAS river analysis system: Supplemental to HEC-RAS version 5.0 user's manual* (Version 5.0.4). Davis, CA: U.S. Army Corps of Engineers Hydrologic Engineering Center.

U.S. Environmental Protection Agency. (2013). *International summit on integrated environmental modeling* (EPA/600/R/12/728). Athens, GA: U.S. Environmental Protection Agency, Office of Research and Development, Ecosystems Research Division.

U.S. Environmental Protection Agency. (2018). Ecosystem services. Retrieved June 2018 from https://www.epa.gov/eco-research/ecosystem-services

Van der Lee, G. E. M., Van der Molen, D. T., Van den Boogaard, H. F. P., & Van der Klis, H. (2006). Uncertainty analysis of a spatial habitat suitability model and implications for ecological management of water bodies. *Landscape Ecology, 21*(7) 1019–1032.

Wellen, C., Kamran-Disfani, A.-R., & Arhonditsis, G. B. (2015). Evaluation of the current state of distributed watershed nutrient water quality modeling. *Environmental Science & Technology*, *49*(6), 3278–3290.

7

Examples of Stormwater, Watershed, and Receiving Water Quality Models

Debabrata Sahoo, Ph.D., P.E, P.H. and John Schooler, Jr., P.E.

This section presents summary information of various models that are divided into the following units as found in Chapter 4:

1. Watershed quality models
2. Urban stormwater quality models
3. Receiving water quality models
4. Water quality compliance modeling
5. Integrated modeling systems

Each model description depicts information such as availability of the model, types of modeling, pollutant types, model components, model inputs, model limitation, and so on. This chapter should be used as a guidance report. For in-depth information, users are requested to refer to the literature cited after each model description and to Chapter 4 of this manual.

1.0 WATERSHED QUALITY MODELS

1.1 Areal Nonpoint Source Watershed Environment Response Simulation (ANSWERS)

1. Availability of the Model

   The ANSWERS model is available through the Department of Biological Systems Engineering at Virginia Tech. The users should be able to use the program; however, currently there is no technical support of the model.

2. Types of Modeling and Potential Application Areas

   The ANSWERS model is an event-based program; it is based on the premise of distributed parameters to model spatially varying hydrologic processes such as runoff, infiltration, subsurface drainage, and erosion in an ungauged watershed. The program is designed primarily for agricultural watersheds. Generally, the program is used to evaluate the effectiveness of best management practices (BMPs) in agricultural and urban watersheds in managing runoff and erosion. The model simulates the combinatorial effects of land uses, land management, and conservation practices on watershed hydrology and erosion. The model consists of a hydrologic component, a sediment transport component, and various routing components that are necessary to describe the flow of water in different hydrologic compartments. The conceptual basis for the hydrologic model was developed by Huggins and Monke (1966). The governing equation for the simulation of erosion is the continuity equation used by Foster and Meyer (1972).

3. Pollutant Types

   The program can simulate runoff and sediment.

4. Model Components, Techniques, and Processes

   The watershed area is divided into multiple grids where grid properties such as land use, soil, slope, nutrients, and crops are considered homogeneous.

5. Input Data Requirements

   The model has a geographic information system (GIS)-based user interface called QUESTIONS that aids the user in selecting topography, textural soil and land-use maps, management description, and breakpoint rainfall and daily mean air temperature and solar radiation information.

6. Simulation Outputs

   The model provides numerical output.

7. Model Limitations

   The datafile needed to run the program is somewhat complex. In the past, the model needed huge computation power.

8. Recommended Reading

   Beasley, D. B., Huggins, L. F., & Monke, E. J. (1980). ANSWERS: A model for watershed planning. *Transactions of the ASAE, 23*(4), 938–944.

Foster, G. R., & Meyer, L. D. (1972). A closed-form soil erosion equation for upland areas. In H. W. Shen (Ed.), *Sedimentation*, 1-19. Fort Collins, CO: Colorado State University.

Huggins, L. F., & Monke, E. J. (1966). The mathematical simulation of the hydrology of small watersheds. Technical Report 1; Water Resources Research Center. West Lafayette, IN: Purdue University. 130, p. 9.

1.2 U.S. Environmental Protection Agency Screening Procedures

1. Availability of the Model

   This method is available from the U.S. Environmental Protection Agency (U.S. EPA).

2. Types of Modeling and Potential Application Areas

   This is an important U.S. EPA-approved qualitative tool to assess various watershed practices before making any recommendation for water quality improvements in agricutlral, rural, and urban watersheds. The tool also assists in quantifying cost-benefit analysis of various management strategies.

3. Pollutant Types

   The program helps towards general watershed improvement strategies including water quality, erosion control, habitat improvements, and ecosystem restoration.

4. Model Components, Techniques, and Processes

   The program has various components such as management options, pollutant reduction efficiencies, and legal requirements that are used as the decision variables to assess the outcome.

5. Input Data Requirements

   Since this is a screening-level method to narrow down the models based on the user's needs, the program doesn't need any input data such as those that are typically required by the watershed models. Model components differ based on the objectives, so the method requires different information to select the appropriate model. The model components talk about the needs.

6. Simulation Outputs

   The program is used as a screening tool.

7. Model Limitations

   The model is applicable only for screening.

8. Recommended Reading

   U.S. Environmental Protection Agency. (1985). *Water quality assessment: A screening procedure for toxic and conventional pollutants in surface and ground water* (EPA/600/6-85/002a). Athens, GA: Environmental Research Laboratory.

1.3 Regression Method

1. Availability of the Model

   This is a method. No specific computer model is available.

2. Types of Modeling and Potential Application Areas

   The regression method is applied to establish the statistical relationship between multiple variables in different types of watershed studies. Depending on the problem, regression modeling can be used to manage watersheds, water quality, or climate change studies. This is an important U.S. EPA-approved qualitative tool to assess various watershed practices before making recommendations for water quality improvements. The tool also assists in quantifying cost-benefit analysis of various management strategies.

3. Pollutant Types

   In general, the method can be applied to any pollutant wherein a relationship with another variable is possible.

4. Model Components, Techniques, and Processes

   The model building needs sufficient data to build a statistical relationship that would work within the boundary condition(s).

5. Input Data Requirements

   In general, users must insert data that are required to establish the relationship. For example, flow and turbidity offer a great relationship if the user wants to quantify sediment loadings.

6. Simulation Outputs

   Predicted variable.

7. Model Limitations

   This method can only be used for a certain set of conditions within the boundary conditions of the data used.

8. Recommending Reading

   Galloway, J. M. (2014). Continuous water-quality monitoring and regression analysis to estimate constituent concentrations and loads in the Red River of the North at Fargo and Grand Forks,

North Dakota, 2003-12. U.S. Geological Survey Scientific Investigations Report 2014–5064. Retrieved from https://pubs.er.usgs.gov/publication/sir20145064

1.4 Simple Method

1. Availability of the Model

   This method is widely available for use without cost and is not a specific computer program. The original method was described in Thomas Schueler's 1987 publication, *Controlling urban runoff: A practical manual for planning and designing urban BMPs* (Schueler & Metropolitan Washington Water Resources Planning Board, 1987). Although there is no support available for this method, there are plenty of literature and studies.

2. Types of Modeling and Potential Application Areas

   As the name of the model implies, the Simple Method quickly helps end users in estimating pollutant loads from a variety of urban land uses. The model uses pollutant concentrations, drainage area, impervious area, and annual precipitation to estimate the loads. The method can also be used as a planning-level tool.

3. Pollutant Types

   In general, the method can be applied to any pollutant for which loading is to be estimated.

4. Model Components, Techniques, and Processes

   The model uses mass balance relationship to compute pollutant loadings.

5. Input Data Requirements

   In general, datasets such as precipitation, land use, drainage area, and event mean concentration are needed to compute loadings.

6. Simulation Outputs

   The numeric value of the pollutant loadings.

7. Model Limitations

   The model uses a simple mass balance approach to compute loadings. The numbers generated should be used for planning-level purposes.

8. Recommended Reading

Schueler, T. R., & Metropolitan Washington Water Resources Planning Board. (1987). *Controlling urban runoff: A practical manual for planning and designing urban BMPs* (Publ. No. 87703).

Washington, D.C.: Metropolitan Washington Council of Governments.

1.5 Watershed Analysis Risk Management Framework (WARMF)

1. Availability of the Model

   As a public domain model created and supported by a private firm, WARMF is available without cost from Systech Water Resources, Inc. Systech Water Resources has information available on their website (https://systechwater.com/warmf_software/) as well as an online help system (https://warmf.com/home/). Several documents and peer reviews are also available on Systech Water Resources' website.

2. Types of Modeling and Potential Application Areas

   The WARMF is a distributed-parameter model developed and supported by SYSTECH Water Resources that is used to simulate hydrology and pollutants to aid in modeling watersheds, assessing water quality impairment, and understanding the impact of climate change and real-time management. The model can simulate up to 40 water quality constituents. The program has a bioenergetics model that quantifies bioaccumulation of constituents in the food chain, which is a unique attribute of the program.

3. Pollutant Types

   The program can simulate runoff, sediment, and chemicals such as nitrogen and phosphorus, ions, organic carbon, dissolved oxygen, pH, phytoplankton, pesticides, mercury, arsenic, selenium, and heavy metals.

4. Model Components, Techniques, and Processes

   The program is a distributed parameter model that simulates the rainfall-runoff process and fate and transport of approximately 40 water quality constituents.

5. Input Data Requirements

   The program has a good user interface and easy access to inputs and outputs. After dividing the model into hydrologic catchments, river segments, and lake segments, inputs include land use classifications, canopy, surface, and up to five soil layers.

6. Simulation Outputs

   The program can simulate in daily or shorter time steps. The program has an interface to visualize output results. The model can be simulated in timesteps of 1 minute to 24 hours.

7. Model Limitations

   The model has a non-GIS-based graphical user interface (GUI). The user has to estimate all the model inputs externally and feed the inputs to the model in the GUI.

8. Recommended Reading

   Chen, C. W., Herr, J., & Ziemelis, L. (1998). *Watershed Analysis Risk Management Framework – A decision support system for watershed approach and TMDL calculation* (Documentation Report TR110709). Palo Alto, CA: Electric Power Research Institute.

   Systech Water Resources, Inc. (2018). WARMF Software. Retrieved from https://systechwater.com/warmf_software/

1.6 Agricultural Non-Point Source Pollution Modeling System (AGNPS)

1. Availability of the Model

   Developed by the United States Department of Agriculture (USDA) National Resources Conservation Service (NRCS), information on AGNPS is available on NRCS's website (https://go.usa.gov/KFO). The AGNPS was originally developed as a single-event model, but now refers to a system of modeling components. This model is available for download from the NRCS website without cost. Support is limited to co-operating users who register with the NRCS.

2. Types of Modeling and Potential Application Areas

   The original single event AGNPS model has been discontinued since the mid 1990s. The current usage of AGNPS refers to a system of modeling components. The program was modified to enhance the functionalities of the program, automate various inputs, and assess the outputs. The AGNPS is an event-based model that is used to assess the effect of watershed management strategies, particularly in agricultural watersheds affecting water, sediment, and various chemical loadings.

3. Pollutant Types

   The program can simulate runoff, sediment, and chemicals such as nitrogen, phosphorus, organic carbon, and pesticides.

4. Model Components, Techniques, and Processes

   The program is a rainfall-runoff model with distributed parameters. The model is designed to operate on a uniform rectangular grid, with the assumption that all the model characteristics are homogeneous within the grid.

5. Input Data Requirements

   The model requires 22 input parameters that are derived from three input datasets (land cover, digital elevation model, and soil image). Data is input through a GIS-assisted program and an input editor tool. Topographic data, soils, climate/weather, land management practices, and, if desired, ephemeral gully information.

6. Simulation Outputs

   The model provides numerical output that can be used with the Summarization Tool to Evaluate AnnAGNPS Data (STED).

7. Model Limitations

   The model's requirement of 22 input parameters makes the input file complex. The model is designed for applications in agricultural watersheds.

8. Recommended Reading

   Bingner, R. L., & Theurer, F. D. (2016). AGNPS website. Retrieved from http://www.ars.usda.gov/Research/docs.htm?docid=5199

   USDA Natural Resources Conservation Service. (n.d.). AGNPS Home. United States Department of Agriculture. Retrieved from http://go.usa.gov/KFO

1.7 Generalized Watershed Loading Functions (GWLF)

1. Availability of the Model

   The current distributed computer model was made specifically for Pennsylvania, but it has been adopted for uses in the northeastern United States. In addition to a user's manual, guidance documents, and several videos, support is provided by Pennsylvania State University (PSU). The software is available for download without cost from PSU's website.

2. Types of Modeling and Potential Application Areas

   The GWLF model is a combined distributed/lumped parameter watershed model that is used for medium-sized watershed modeling to assess the impact from non-point sources in urban and rural areas. The model can also simulate loads from the septic systems and point sources. In general, the model can simulate streamflow, watershed erosion and sediment yield, total nitrogen and phosphorus loads in

streamflow, erosion, total nitrogen, and total phosphorus loading from each land use.

3. Pollutant Types

   The program can simulate runoff, sediment yield, bacteria, total nitrogen, and total phosphorus.

4. Model Components, Techniques, and Processes

   The program describes non-point sources runoff and erosion using the distributed modeling concept. It uses the lumped parameter linear reservoir groundwater model. Point sources are summed by constant mass loads.

5. Input Data Requirements

   The GWLF model requires daily precipitation and temperature data. The program also requires runoff sources, transport, and chemical parameters.

6. Simulation Outputs

   The users can obtain the model outputs in tabular and graphical form.

7. Model Limitations

   In general, the model has simplified stream transport and water quality simulations.

8. Recommended Reading

Haith, D. A., Mandel, R., & Wu, R. S. (1992). *GWLF: Generalized Watershed Loading Functions* (Version 2.0). Department of Agricultural & Biological Engineering. Retrieved from www.mapshed.psu.edu/Downloads/GWLFManual.pdf

University of California, Davis. (n.d.). GWLF: Generalized Watershed Loading Function. Retrieved from http://cwam.ucdavis.edu/pdfs/GWLF.pdf

1.8 Hydrologic Simulation Program in FORTRAN (HSPF)/Loading Simulation Program in C++ (LSPC)

1. Availability of the Model

   This model is sponsored by U.S. EPA and USGS and is available for download from the USGS website (https://water.usgs.gov/software/HSPF/). AQUA TERRA Consultants provide the model updates and support services (http://www.aquaterra.com/resources/hspfsupport/) in addition to a webpage to download HSPF and several ancillary programs, all for free.

2. Types of Modeling and Potential Application Areas

   The HSPF model is one of the first comprehensive watershed models developed in the United States that performs watershed hydrology, instream hydraulics, and fate and transport of pollutants. It is a continuous simulation model. The program runs within the BASINS software. The program is used for both water quantity and quality modeling, particularly to address total maximum daily load (TMDL) issues. The program is vastly used for urban watersheds. The LSPC is a program that is very similar to HSPF, and it is programmed in an updated environment to make the model efficient and user friendly.

3. Pollutant Types

   The program can simulate runoff, interflow, base flow, snowmelt, dissolved oxygen, biochemical oxygen demand, temperature, pesticides, sediment detachment and transport, pH, species of nitrogen and phosphorus, phytoplankton, zooplankton, and so on.

4. Model Components, Techniques, and Processes

   The program is a semi-distributed, lumped parameter model. The model simulates the process in three modules: PERLND, IMPLAND, and RCHES.

5. Input Data Requirements

   Watershed physical measurements, such as the physical properties of a stream reach, land use, and soil, are required to characterize the watershed in the model. Continuous meteorological data such as precipitation and potential evapotranspiration are required for the watershed simulation. Air temperature, dew point temperature, wind, and solar radiation are required for snowmelt simulations. Datasets such as wind speed, solar radiation, humidity, cloud cover, tillage practice, and point source are needed for water quality simulations.

6. Simulation Outputs

   The output of the model runs can be visualized using external programs or by using tools provided in BASINS.

7. Model Limitations

   The model by itself is a non-GIS program. It is a very data-intensive model and requires significant user expertise.

8. Recommended Reading

   USGS. (n.d.). Summary of HSPF. *U.S. Geological Survey*. Retrieved from https://water.usgs.gov/cgi-bin/man_wrdapp?hspf

1.9 Soil and Water Assessment Tool (SWAT)

1. Availability of the Model

   This is a public domain model jointly developed by the USDA Agricultural Research Service (USDA-ARS) and Texas A&M AgriLife Research, part of The Texas A&M University System. The SWAT model and numerous add-on programs are available for download from the SWAT website (https://swat.tamu.edu/) at no cost. Support is provided by means of new user workshops, publications, an online support group, and contact with the developers. There are regular conferences for SWAT users to gather and present projects and ideas.

2. Types of Modeling and Potential Application Areas

   The SWAT model is a physically based computer program developed by the USDA to assess the impact of land management strategies in a watershed, particularly the impact from agricultural practices. The main components of SWAT include weather inputs, surface runoff, return flow, infiltration, evapotranspiration, transmission losses, pond and reservoir storage, crop growth and irrigation, groundwater flow, reach routing, nutrient and pesticide loading, and water transfer. Because the model is a physically based model, the program can be used to simulate hydrology and water quality in ungauged watersheds. The program can be used for long-term hydrologic and water quality simulations. The program has a GIS version and a non-GIS version. The model requires daily measured precipitation inputs, and the outputs are reported in daily time step.

3. Pollutant Types

   The program can simulate runoff, sediment, nutrients, and other agricultural pollutants.

4. Model Components, Techniques, and Processes

   The program is a physically based model. The land area in the subbasin is divided into hydrologic responses units (HRUs), which are based on the unique land use, management, and soil attributes. The HRUs represent the smallest unit in the program with which the computations are carried. The program uses the NRCS curve number method to compute runoff. Sub-daily precipitation data are needed if the Green and Ampt method is selected. The pollutants are routed through the stream. Channel water routing uses the following two methods: the variable storage method and the Muskingum method.

5. Input Data Requirements

   The SWAT model requires topographic, climate, soil, land use, and management input parameters.

6. Simulation Outputs

   The SWAT model has components to read the outputs from the model and display the output in a graphical form. Several levels of outputs, such as average daily flows, average daily loads, and average monthly flows, can be generated from the model.

7. Model Limitations

   The program is less intuitive, highly complex, and very data intensive. It requires significant hydrology and water quality knowledge. The program is less applicable to perform storm analysis.

8. Recommended Reading

   Texas A&M University, Texas A&M AgriLife Research, & USDA. (2019). Soil & Water Assessment Tool. Retrieved from https://swat.tamu.edu/

1.10 Water Erosion Prediction Project (WEPP)

1. Availability of the Model

   The current version (v2012.8) is available for download without cost from the USDA-ARS website (https://www.ars.usda.gov/midwest-area/west-lafayette-in/national-soil-erosion-research/docs/wepp/). Model documentation is provided on the website, but no information is given for technical support.

2. Types of Modeling and Potential Application Areas

   The WEPP model is a process-based, distributed parameters, continuous simulation model. The model is applicable to hillslopes and to smaller watersheds to model soil erosion. The distributed input parameters include rainfall, soil, plant growth parameters, effects of tillage, slope shape, steepness, soil erodibility parameters, and residue decomposition parameters.

3. Pollutant Types

   The program can simulate runoff and soil erosion.

4. Model Components, Techniques, and Processes

   The program, when applied to hillslopes, can be divided into nine conceptual components. The WEPP simulations are generated using the CLIGEN model. Winter processes include soil frost and thaw development, snowfall, and snow melting. The hydrology computes infiltration, runoff, evaporation, transpiration, percolation, interception, and depressional storage. The impacts from tillage are computed in the soil compartment. For more detailed information, refer to the users' document.

5. Input Data Requirements

   Climate data file, a slope file, soil data, and cropping/management file.

6. Simulation Outputs

   Generally, the model outputs runoff and erosion storm-by-storm information and monthly and annual summary statistics. The output has both the temporal and spatial outputs.

7. Model Limitations

   The program should not be used for large-scale watershed simulations. Erosion predictions are applicable to only field sizes.

8. Recommended Reading

   USDA. (1995). *WEPP user summary: USDA-Water Erosion Prediction Project* (NSERL Report NO. 11). Dennis C. Flanagan & Stanley J. Livingston (Eds.). West Lafayette, IN: National Soil Erosion Research Laboratory. Retrieved from https://www.ars.usda.gov/ARSUserFiles/50201000/WEPP/usersum.pdf

2.0 URBAN STORMWATER QUALITY MODELS

2.1 Program for Predicting Polluting Particle Passage Through Pits, Puddles, and Ponds–Urban Catchment Model (UCM)

1. Availability of the Model

   This model was first created in 1990, and it is currently on Version 3.5 (completed in 2015). The model, documentation, and other resources are available at https://www.wwwalker.net/p8/. Technical support is available for a fee.

2. Types of Modeling and Potential Application Areas

   The P8-UCM model is a physically based stormwater quality model that simulates the generation and transport of pollutants in an urban watershed. The program (Version 2.1) simulates runoff and pollutant transport for a maximum of 192 watersheds, 48 low-impact development (LID) BMPs, five particle sizes, and 10 water quality components. The model can simulate various LID BMPs such as infiltration basins, swales, buffer strips, dry pond, wet pond, extended pond, and flow splitters. The program is used by engineers, planners, and practitioners to evaluate the impact of development with or without the implementation of the LID BMPs in the watershed. It has been used in the regulatory world in states such as Minnesota, New Mexico, and Wisconsin.

3. Pollutant Types

   The model can predict transport and trapping of suspended solids, total phosphorus, total Kjeldahl nitrogen, copper, lead, zinc, and total hydrocarbons. The program can simulate runoff, sediment, nutrients, and other pollutants.

4. Model Components, Techniques, and Processes

   Model simulations are performed by feeding the program with continuous hourly rainfall and daily air temperature time series. Runoff from pervious areas is simulated using the SCS curve number method; however, runoff from impervious areas is generated after the cumulative rainfall exceeds the depression storage. Infiltration in pervious areas is quantified by computing the difference between rainfall and runoff plus evapotranspiration, as expressed by the equation: Infiltration = rainfall − (runoff + evapotranspiration). Evapotranspiration is computed from air temperature and season according to Hamon's method. The model uses a SWMM-like sediment rating model for pollutant concentrations.

5. Input Data Requirements

   Hourly precipitation, daily temperatures, BMP characteristics, watershed data, water quality parameters, evapotranspiration, and snowmelt.

6. Simulation Outputs

   The model provides tabular and graphical outputs for precipitation, flow, and pollutants. The program also provides data in ASCII format to be used later in Word- or Excel-type programs.

7. Model Limitations

   The program cannot perform instream water quality simulations. It is limited by the number of watersheds. The program has three water quality BMPs (swale/buffer, infiltration basin, detention pond).

8. Recommended Reading

   Walker, W. W., Jr., & Walker, J. D. (2017). P8 Urban Catchment Model. Retrieved from http://www.wwwalker.net/p8/

2.2 Source Loading and Management Model (SLAMM)/ WinSLAMM

1. Availability of the Model

   The software for this model is available for purchase through PV & Associates, LLC at www.winslamm.com. The current version is

10.4.1, which was released in March 2019. The PV & Associates website includes documentation, information on training opportunities, and instructions for purchasing the software. Limited support is provided to licensees.

2. Types of Modeling and Potential Application Areas

   The WinSLAMM model is designed to solve urban stormwater quality issues. The model evaluates runoff volume and loading from the individual source area for every rainfall event. The model does not lump the land uses together. It estimates the hydrology and water quality from each land use category. The model can be used to address TMDL implementation plans and urban stormwater management strategies and determine optimal LID strategies among other urban water quality issues.

3. Pollutant Types

   The model can simulate runoff, particulate solids, chemicals such as total phosphorus and total nitrogen, and sediment.

4. Model Components, Techniques, and Processes

   The program has various objects in the software interface. Objects such as residential area and outfall can be dragged to the user interface window before running the model. The program describes nonpoint sources runoff and erosion using the distributed modeling concept. It uses a lumped parameter linear reservoir groundwater model. Point sources are summed by constant mass loads.

5. Input Data Requirements

   The WinSLAMM requires rainfall, wash-off coefficients, runoff coefficients, and particulate solids concentration. The program also requires runoff sources, transport, and chemical parameters.

6. Simulation Outputs

   Users can access outputs in tabular format and also in graphical format. Users can also obtain flow duration curves.

7. Model Limitations

   The model is applicable to the urban system.

8. Recommended Reading

PV & Associates, LLC. (2019). WinSLAMM Select Documentation. Retrieved from http://www.winslamm.com/Select_documentation.html

2.3 Stormwater Management Model (SWMM)

1. Availability of the Model

   This model was originally developed in 1977 and has been upgraded numerous times since then. The most recent version is 5.1.013, released in 2018. The program can be downloaded for free on the U.S. EPA website (https://www.epa.gov/water-research/storm-water-management-model-swmm). A user manual, documentation, and other resources are also available on the U.S. EPA website. Technical support may be limited (an email address is provided on the Website). A listserv hosted by the University of Guelph is also available.

2. Types of Modeling and Potential Application Areas

   The SWMM is a single-event and a long-term simulations model for water quantity and quality that is primarily used in urban areas. The processes are physically based. The program has hydraulic modeling capabilities. The user can estimate pollutant loadings using this program. The user can also model LID and green infrastructure solutions using this program. The SWMM has a tool for climate projections. There are several programs based on the SWMM engine such as XPSWMM, PCSWMM, and InfoSWMM. Some of these SWMM models are capable of conducting two-dimensional modeling.

3. Pollutant Types

   The program can simulate runoff and pollutants such as nitrogen, phosphorus, sediment, bacteria, pesticides, and metals. The user needs to supply the pollutant buildup and wash-off rate.

4. Model Components, Techniques, and Processes

   The program is a link-node type model that can compute hydrology in pervious and impervious layers. The model is provided with time series of rainfall information or design storm, various parameters to estimate runoff, and parameters to estimate water quality, among other inputs. The model has five infiltration processes.

5. Input Data Requirements

   The model requires various hydrologic parameters such as watershed area, curve number, rainfall information, soil information, and infiltration information. The model also requires various water quality information such as pollutant loading rates and wash-off rates.

6. Simulation Outputs

   The SWMM outputs can be viewed in numerous ways. The user can view the results graphically or can export the results in a data format, which can be viewed using external programs.

7. Model Limitations

   The program is primarily used in urban areas and is not suitable for the rural areas.

8. Recommended Reading

   U.S. Environmental Protection Agency [U.S. EPA]. (2015). Storm Water Management Model user's manual Version 5.1. U.S. *EPA National Service Center for Environmental Publications (NSCEP)*. Retrieved from https://nepis.epa.gov/Exe/ZyPURL.cgi?Dockey=P100N3J6.txt

2.4 National Stormwater Calculator

1. Availability of the Model

   The program was developed and distributed by U.S. EPA. The program is available on the U.S. EPA website (https://www.epa.gov/water-research/national-stormwater-calculator). The program can assist non-technical professionals in performing screening-level analysis of stormwater runoff for small- to medium-sized sites (less than 48 500 square meters [12 acres]). A user's guide is available on the website, but support may be limited (email address provided on website).

2. Types of Modeling and Potential Application Areas

   The program can simulate stormwater runoff generated from a site under different land use change and watershed developmental conditions. It can also simulate runoff from different LID BMPs. Generally, this program can be used to assess the reduction in runoff due to different BMP implementation along with the relative cost of implementing various LID BMPs. The program can be used to simulate runoff for future climate change conditions. The following BMPs are available in the current version: disconnections, rain harvesting, rain gardens, green roofs, street planters, infiltration basins, and permeable pavement. The tool has been applied in Baltimore, Maryland, to assess the flow reduction and planning-level costs of the BMPs.

3. Pollutant Types

   The program can only simulate runoff.

4. Model Components, Techniques, and Processes

   The program is based on the SWMM 5 dynamic rainfall-runoff simulation model for long-term simulation of runoff quantity. The program uses a nonlinear reservoir model to estimate surface runoff produced from the land and Green-Ampt method for infiltration.

5. Input Data Requirements

   Hydrologic characteristics of the drainage area(s) and BMP characteristics.

6. Simulation Outputs

   The model provides numerical and graphical outputs for rainfall and flow.

7. Model Limitations

   The software can only be implemented at site level (less than 12 ac). It does not have water quality simulation capabilities. The program cannot perform instream water quality simulations. The software relies on online resources and available databases, which limits the user.

8. Recommended Reading

Rossman, L. A., & Bernagros, J. T. (2019). National stormwater calculator user's guide—version 1.2.0.3. U.S. EPA. Retrieved from https://www.epa.gov/sites/production/files/2019-04/documents/swc_users_guide_desktop_v1.2.0.3_april_2019.pdf

U.S. Environmental Protection Agency. (2019). National Stormwater Calculator. *U.S. EPA*. Retrieved from https://www.epa.gov/water-research/national-stormwater-calculator

2.5 Watershed Treatment Model (WTM)

1. Availability of the Model

   This model is available for download without cost from the Center for Watershed Protection on their website (https://owl.cwp.org/mdocs-posts/watershed-treatment-model-wtm-2013/). The most current version, updated in 2018, is Version 1.0.2. Documentation and a user's guide are also available online, but support is not readily available.

2. Types of Modeling and Potential Application Areas

   The WTM is a simple screening-level spreadsheet tool that assists the user in evaluating loads from a wide variety of sources and assists in assessing treatment alternatives. The program can be used as a starting point in developing watershed management strategies.

3. Pollutant Types

   The program can simulate runoff, sediment, nutrients, and bacteria on an annual basis.

4. Model Components, Techniques, and Processes

   The program has three components: pollutant sources, treatment options, and future growth. The first component identifies the pollutant sources and computes the loading from the sources, the second component evaluates the reduction in pollutant loads, and the third component takes into account the future growth in the watershed assessment process.

5. Input Data Requirements

   The model requires watershed drainage area, impervious cover, stormwater runoff pollutant concentrations, annual precipitation and runoff coefficiencts, loads from secondary sources such as SSOs, treatment efficiencies of the BMPs, and effectiveness factors for education.

6. Simulation Outputs

   The simulations split pollutant source load, the benefit of the existing practice, benefit of the future practice, and the effects of growth, all on an annual basis.

   The WTM has tools to read the outputs from the model and display the output in a graphical form. Several levels of outputs, such as average daily flows, average daily loads, and average monthly flows, can be generated from the model.

7. Model Limitations

   The program is useful for urban areas. The results should be used cautiously and should not be used for regulatory and compliance purposes.

8. Recommended Reading

   Caraco, D. (2013). Watershed Treatment Model (WTM) 2013 documentation. *Center for Watershed Protection, Inc.* Retrieved from https://owl.cwp.org/?mdocs-file=5494

3.0 RECEIVING WATER QUALITY MODELS

3.1 Hydrologic Engineering Center River Analysis System (HEC-RAS)

1. Availability of the Model

   The HEC-RAS is distributed by USACE and is available without cost on a dedicated website (http://www.hec.usace.army.mil/software/hec-ras/). The website includes documentation, user guides, and bug reporting; however, active technical support is only available to USACE customers.

2. Types of Modeling and Potential Application Areas

   The HEC-RAS model is a physically based model designed to simulate steady and unsteady flows. The program is designed to perform one- and two-dimensional hydraulic calculations for natural and constructed channels such as river flow simulations, bridge scour analysis, and channel hydraulic design.

3. Pollutant Types

   The program can simulate flow depth and water quality components such as temperature, dissolved nitrogen, dissolved phosphorus, algae, dissolved oxygen, and carbonaceous biochemical oxygen demand (CBOD).

4. Model Components, Techniques, and Processes

   The program is a physically based model. The channel is divided at multiple points based on observations, channel characteristics, and flow behavior.

5. Input Data Requirements

   The model requires flow information, Manning's value, and geometric data.

6. Simulation Outputs

   The model can provide steady flow water surface profiles, one- and two-dimensional unsteady flow simulations, sediment, and water quality. The program has graphics that can be used to visualize the outputs.

7. Model Limitations

   The program has simplified instream water quality process routines.

8. Recommended Reading

   U.S. Army Corps of Engineers. (2016). HEC-RAS 5.0 users manual. *USACE Hydraulic Engineering Center*. Retrieved from http://www.hec.usace.army.mil/software/hec-ras/documentation/HEC-RAS%205.0%20Users%20Manual.pdf

3.2 QUAL2E

1. Availability of the Model

   This model was developed by U.S. EPA and is now available for free as a part of the BASINS modeling suite. User information and guidance is available in the form of a user manual, tutorials, training, technical notes, and publications. These are available on the BASINS page on the U.S. EPA website (https://www.epa.gov/ceam/better-assessment-science-integrating-point-and-non-point-sources-basins).

2. Types of Modeling and Potential Application Areas

   The QUAL2E is a U.S. EPA-approved, steady state, one-dimensional model to assess the fate and transport of general pollutants in streams, rivers, and lakes. The model can be used as a planning-level tool for water quality improvements from waste loads. The program has the functionalities to model the meteorological effect on water quality.

3. Pollutant Types

   The program can model temperature, chlorophyll-a, organic nitrogen, ammonia, nitrate, nitrite, organic phosphorus, dissolved phosphorus, CBOD, dissolved oxygen, coliforms, a non-conservative constituent, and three conservative constituents.

4. Model Components, Techniques, and Processes

   The model network is divided into three parts: headwaters, reaches, and junctions. Mass transport in the programs is handled in a simpler manner.

5. Input Data Requirements

   The program needs hydraulic characteristics, flow and water quality information, and climatological information for algae and temperature simulations.

6. Simulation Outputs

   Simulation results provide information on hydraulics, reaction coefficient, and water quality. The summary table contains flows, velocities, travel time, depths, and cross-sectional area.

7. Model Limitations

   The current program can model to a maximum of 50 allowable reaches with 20 computational elements in each reach, a maximum of 10 headwater elements, a maximum of nine junction elements, and a maximum of 50 point-source and withdrawal elements. The program is well suited for point-source-dominated stream.

8. Recommended Reading

   Brown, L., & Barnwell, T. (1987). The enhanced stream water quality models QUAL2E and QUAL2E-UNCAS: Documentation and user manual. Environmental Protection Agency.

3.3 Water Quality Analysis Simulation Program (WASP)

1. Availability of the Model

   The current WASP model represents an enhancement of the original model. The current version, WASP 8.32, is available for download without cost from U.S. EPA (https://www.epa.gov/ceam/water-quality-analysis-simulation-program-wasp). Frequently asked questions, documentation, tutorials, example files, and an email support group are all available on this website.

2. Types of Modeling and Potential Application Areas

   The WASP is a dynamic multi-dimensional model to simulate fate and transport of various contaminants in surface waters. The program can be used to understand the impact of point and non-point sources on instream water quality. The model can simulate most of the water quality process in a mechanistic way. The program can perform simulations in one, two, and three dimensions. The model can be coupled with external watershed models. The model is used to understand eutrophication problems by using the advanced eutrophication module. The recent version has a sediment digenesis module.

3. Pollutant Types

   Many traditional and non-traditional pollutants can be simulated using this program, including temperature, dissolved oxygen, nitrogen, phosphorus, sediment, bacteria, and metals. The program is basically used to address eutrophication issues.

4. Model Components, Techniques, and Processes

   The program has advanced eutrophication and advanced toxicant modules. It uses the water resources database (WRDB) to store and manage datasets. The model can simulate advection, dispersion, point loadings, and mass loadings.

5. Input Data Requirements

   The model requires loading and flows information from the point and non-point sources, channel geometry to estimate the hydrodynamics, weather information to build time functions, and algae information to simulate algal conditions.

6. Simulation Outputs

   The outputs can be obtained in days for the pollutant of interest. The minimum allowable output time step is 0.00010 days (or roughly 8.64 seconds). The outputs can be viewed using the BMDUtil and WRDB graphs.

7. Model Limitations

   In general, the model needs enormous amount of in-stream information such as solar radiation, cloud cover, and sediment digenesis. The model depends on the watershed load, which is obtained from another model such as LSPC.

8. Recommended Reading

   U.S. Environmental Protection Agency. (n.d.). Water Quality Analysis Simulation Program (WASP). Retrieved from https://www.epa.gov/ceam/water-quality-analysis-simulation-program-wasp

   WASP8 Development Team. (n.d.). Water Quality Analysis Simulation Program (WASP) User Community Supported Website. *WASP8 Development Team*. Retrieved from http://epawasp.twool.com/

3.4 TOXMOD

1. Availability of the Model

   Very few online resources are available for this model. The model was developed by North American Lake Management Society (NALMS). The program is available by ordering to NALMS, 1 Progress Boulevard, Box 27, Alachua, Florida, 32615.

2. Types of Modeling and Potential Application Areas

   The TOXMOD is a steady state water quality model and a tool to assess long-term water quality conditions of lakes and reservoirs. The system is presented as a well-mixed reactor over a well-mixed sediment layer.

3. Pollutant Types

   The model simulates most of the toxic compounds and sediment.

4. Model Components, Techniques, and Processes

   The model computes a mass balance of solids and toxics partitioned into dissolved and particulate forms.

5. Input Data Requirements

   The program requires a wide variety of data such as lake depth, surface area, sediment thickness, burial rates, sorption, volatilization coefficients, dissolved organic carbon concentration, time series of flows, inflow concentration, and so on.

6. Simulation Outputs

   The model comes with graphical and tabular output options for sediment and toxicants.

7. Model Limitations

   The model cannot simulate reverse flows.

8. Recommended Reading

Riecken, S. (1995). *A compendium of water quality models.* British Columbia, Canada: Water Quality Branch, Ministry of Environment, Lands and Park. Retrieved from https://www2.gov.bc.ca/assets/gov/environment/air-land-water/water/waterquality/water-quality-reference-documents/wq_ref_models_compendium.pdf

Shoemaker, L. (1997). Compendium of Tools for Watershed Assessment and TMDL Development.

3.5 CE-QUAL-ICM

1. Availability of the Model

   This model was developed by USACE Engineer Research & Development Center (ERDC) Environmental Laboratory and is now available without cost from USACE. Several papers are available on the model, but no user manual, tutorials, or technical support are available. Contact information is provided for a person who can facilitate a download on USACE's website (https://www.erdc.usace.army.mil/Media/Fact-Sheets/Fact-Sheet-Article-View/Article/547416/ce-qual-icm-icm/).

2. Types of Modeling and Potential Application Areas

   The CE-QUAL-ICM is a multi-dimensional, time-variable water quality model suitable to assess the fate and transport of pollutants in lakes, rivers, estuaries, and coastal waters. The program has been applied in the Chesapeake Bay, San Juan Bay Estuary, and St. Johns River.

3. Pollutant Types

   The program is used to assess the fate and transport of nitrogen, phosphorus, carbon, suspended solids, temperature, and salinity.

4. Model Components, Techniques, and Processes

   The model has various sub-models such as the sediment diagenesis, filter feeding benthos, toxicants, submerged aquatic vegetation, and carbonate cycle.

5. Input Data Requirements

   The model needs information, such as flow and volume, which is provided using an input file.

6. Simulation Outputs

   The output of the program is obtained in a tabular format, which can be used for graphical analysis.

7. Model Limitations

   The model does not simulate hydrodynamics and flows. The program can only simulate up to 22 state variables.

8. Recommended Reading

   Cerco, C., & Cole, T. (1993). Three-dimensional eutrophication model of Chesapeake Bay. *Journal of Environmental Engineering*, 119(6), 1006–1025.

   Cerco, C., & Moore, K. (2001). System-wide submerged aquatic vegetation model for Chesapeake Bay. *Estuaries*, *24*(4), 522–534.

   Cerco, C., & Noel, M. (2004). Managing for water clarity in Chesapeake Bay. *Journal of Environmental Engineering, 130*(6), 631–642.

   Cerco, C., Noel, M., & Kim, S-C. (2006). Three-dimensional management model for Lake Washington: (II) Eutrophication modeling and skill assessment. *Journal of Lake and Reservoir Management,* 22(2), 115–131.

   Shoemaker, L. (1997) Compendium of Tools for Watershed Assessment and TMDL Development.

3.6 CE-QUAL-RIV1

1. Availability of the Model

   Various documentation sources are available online and in print, although much of the documentation has fallen out of date and availability may be limited. The most recent version of this program, developed by USACE, is Version 2.0, released in 1995.

2. Types of Modeling and Potential Application Areas

   The CE-QUAL-RIV1 is a one-dimensional, unsteady water quality model designed for regulated rivers and unstratified lakes.

3. Pollutant Types

   Constituents that can be modeled in the program are temperature, dissolved oxygen, CBOD, organic nitrogen, ammonia, nitrate, orthophosphorus, coliform bacteria, dissolved iron, and dissolved manganese. The effects of algae can also be modeled.

4. Model Components, Techniques, and Processes

   The model has two components: hydrodynamics and water quality. The output from the hydrodynamic solution is used to guide the water quality model. The hydrodynamic process uses the Newton–Raphson procedure to solve the nonlinear Saint Venant equation.

5. Input Data Requirements

   Input data required for the model are river geometry, channel roughness, later inflows or withdrawls, and initial and boundary conditions.

6. Simulation Outputs

   Depending on the user's requirements, the model produces various water quality outputs, mostly in tabular form.

7. Model Limitations

   The program cannot be used to simulate multi-dimensional flows.

8. Recommended Reading

   Dortch, M., Schneider, T., Martin, J., Zimmerman, M., & ARMY Engineer Waterways Experiment Station Vicksburg MS Environmental Lab. (1990). CE-QUAL-RIV1: A Dynamic, One-Dimensional (Longitudinal) Water Quality Model for Streams. User's Manual. Retrieved from http://agris.fao.org/agris-search/search.do?recordID=AV20120162750

   Environmental Laboratory, U.S. Army Engineer Waterways Experiment Station. (1997). CE-QUAL-RIV1: One-dimensional, dynamic flow and water quality model for streams. Retrieved from https://el.erdc.dren.mil/elmodels/w2info.html

   Shoemaker, L. (1997). Compendium of Tools for Watershed Assessment and TMDL Development.

3.7 River Hydrodynamics Model (RIVMOD-H)

1. Availability of the Model

   This model is only readily available as a component of other models, such as WASP or SWMM. Various documentation and literature are available online and in print, but technical support is no longer available.

2. Types of Modeling and Potential Application Areas

   The RIVMOD-H is a one-dimensional water quality model that can be applied to steady or unsteady systems such as rivers and streams.

3. Pollutant Types

   The program depends on external models for transport processes.

4. Model Components, Techniques, and Processes

   The model uses the equations for momentum and continuity for a one-dimensional, unidirectional flow.

5. Input Data Requirements

   The program requires data such as channel morphometry, bed elevations, and boundary conditions set by the user.

6. Simulation Outputs

   The program produces water surface elevations and unsteady flows at user-defined locations.

7. Model Limitations

   The program has no transport capability and functions only in tandem with an external model such as WASP or SWMM.

8. Recommended Reading

Shoemaker, L. (1997). Compendium of Tools for Watershed Assessment and TMDL Development.

Ward, G. H. Jr., & Benaman, J. (1999). *Models for TMDL application in Texas watercourses: Screening and model review.* (CRWR-99-7). Austin, TX: Center for Research in Water Resources at the University of Texas at Austin. Retrieved from https://repositories.lib.utexas.edu/bitstream/handle/2152/6797/crwr_onlinereport99-7.pdf?sequence=2

3.8 Environmental Fluid Dynamics Code (EFDC)

1. Availability of the Model

   This model is available for download without cost and is supported by U.S. EPA (https://www.epa.gov/ceam/environmental-fluid-dynamics-code-efdc). U.S. EPA's Center for Exposure Assessment Modeling provides technical support for the model. Periodic training may be available, and a four-part user manual and limited documentation are available on the website.

2. Types of Modeling and Potential Application Areas

   The EFDC is a three-dimensional, state-of-the-art model to assess the fate and transport of pollutants, particularly in complex surface water systems including the hydrodynamics of coastal systems. The tool is used for watershed planning and TMDL developments.

3. Pollutant Types

   The program is used for simulating water quality including eutrophication, sediment digenesis, sediment transport, and heat budget calculations. Various pollutants simulated in the program are temperature, dissolved oxygen, nitrogen, phosphorus, sediment, bacteria, pesticide, and metals.

4. Model Components, Techniques, and Processes

   The model has a hydrodynamic component, uses the finite difference method, and models various aspects of sediment transport.

5. Input Data Requirements

   To simulate sediment transport, the model requires information such as initial sediment in fluid phase, initial sediment mass per unit area of bottom surface, sediment-specific volume, sediment-specific gravity, and sediment settling velocity.

6. Simulation Outputs

   The program results in two- and three-dimensional graphics output and visualization.

7. Model Limitations

   The model depends on a watershed model for water quality loading information. The model is a very data intensive model, requiring significant user's expertise.

8. Recommended Reading

   Tetra Tech, Inc. (2007). The Environmental Fluid Dynamics Code user manual U.S. EPA Version 1.01. *Tetra Tech, Inc.* Retrieved from https://www.epa.gov/sites/production/files/2016-01/documents/efdc_user_manual_epa_ver-101.pdf

3.9 CE-QUAL-W2

1. Availability of the Model

   The Water Quality Research Group at the Department of Civil and Environmental Engineering and the Maseeh College of Engineering and Computer Science, Portland State University, maintain the model code and provide model enhancements for the USACE Waterways Experiments Station. This group hosts an annual workshop/training session and also provides documentation, example project summaries, and a user manual on their website (http://www.cee.pdx.edu/w2/). Version 4.1 (released 2018) is available for download from this website without cost.

2. Types of Modeling and Potential Application Areas

   The program can be used to simulate hydrodynamics and water quality of streams, rivers, lakes, reservoirs, and estuaries. The program has two-dimensional longitudinal-vertical simulation capabilities. It is used for systems with distinct longitudinal variations.

3. Pollutant Types

   The program can simulate general water quality parameters such as dissolved oxygen, nutrients, temperature, and sediment.

4. Model Components, Techniques, and Processes

   The model has a hydrodynamic component and a water quality component. The water quality components are modular, allowing the user to add additional subroutines as needed.

5. Input Data Requirements

   The model requires geometric data, hydraulic parameters, calibration parameters, initial conditions, and boundary conditions.

6. Simulation Outputs

   Simulation outputs are obtained in a tabular format and the program lacks modern visualization capabilities.

7. Model Limitations

   The model has limited capability to simulate detailed sediment transport. Model setup and execution are not so user friendly. Lateral variations in velocities, temperature, and water quality constituents are negligible. Water quality interactions are simplified.

8. Recommended Reading

   Cole, T., & Wells, S. (2017). CE-QUAL-W2: A Two-Dimensional, Laterally Averaged, Hydrodynamic and Water Quality Model (Version 4.0). Portland, OR: Department of Civil and Environmental Engineering, Portland State University. Retrieved from http://reviewboard.ca/upload/project_document/EA1314-01_CE-QUAL-W2_2D_Laterally_Averaged_Hydrodynamic__and_WQ_Model_-_User_Manual.PDF

3.10 Cornell Mixing Zone Model (CORMIX)

1. Availability of the Model

   The CORMIX v11.0 model is available in several packages from MixZon, Inc., on their website (http://www.cormix.info/index.php). The free release includes an online user guide and a PDF user's manual, but technical support is only available with a paid subscription. Training courses are offered periodically, also for a fee.

2. Types of Modeling and Potential Application Areas

   The CORMIX is a hydrodynamic model that predicts the fate and transport of the flume from the point-source discharge. It is used for mixing-zone analysis and outfall design.

3. Pollutant Types

   Conservative, non-conservative, heated, brine, or suspended solids can be modeled using this program. Some of the pollutants that can be simulated using this program are temperature, dissolved oxygen, sediment, bacteria, and metals.

4. Model Components, Techniques, and Processes

   The program has a hydrodynamic compartment, visualization routines, and outfall design tools.

5. Input Data Requirements

   The program requires flow characteristics and ambient data.

6. Simulation Outputs

   Model simulations can be viewed using a visualization program.

7. Model Limitations

   The program is limited to point sources or outfall discharges.

8. Recommended Reading

   CORMIX. CORMIX Home Page. Retrieved from http://www.cormix.info/

3.11 BATHTUB

1. Availability of the Model

   Version 6.1 is available for download from the USACE Waterways Experiment Station. The developer, William W. Walker, Jr., Ph.D., maintains documentation, references, and a user's manual on his website (http://www.wwwalker.net/bathtub/help/bathtubWebMain.html). Limited technical support is available by email from a USACE ERDC employee.

2. Types of Modeling and Potential Application Areas

   The BATHTUB is a steady state, one-dimensional model that is used to simulate water quantity and quality in reservoirs and lakes. It is used particularly for eutrophication analysis.

3. Pollutant Types

   The program can be used to model general nutrients (nitrogen and phosphorus) and eutrophication.

4. Model Components, Techniques, and Processes

   The program has three components: FLUX, for analysis of tributary loads into a reservoir; PROFILE, for analysis of reservoir stratification and water quality structure; and BATHTUB, for empirical determination of eutrophication response to pool nutrients.

5. Input Data Requirements

   The program requires information such as inflows, discharges, evaporation, precipitation, kinetic information, loading from the modeled segments, watershed characteristics, and morphology.

6. Simulation Outputs

   The results are obtained in both tabular and graphical forms.

7. Model Limitations

   Up to 40 segments can be modeled using this program.

8. Recommended Reading

   Ward, G. H. Jr., & Benaman, J. (1999). *Models for TMDL application in Texas watercourses: Screening and model review*. (CRWR-99-7). Austin, TX: Center for Research in Water Resources at the University of Texas at Austin. Retrieved from https://repositories.lib.utexas.edu/bitstream/handle/2152/6797/crwr_onlinereport99-7.pdf?sequence=2

3.12 Quality Simulation Along River Systems (QUASAR)

1. Availability of the Model

   This program was developed by the Institute of Hydrology, England. The next version of QUASAR, PC-QUASAR, is available online and can be downloaded from the Center of Ecology and Hydrology (https://www.ceh.ac.uk/services/pc-quasar).

2. Types of Modeling and Potential Application Areas

   The program is a simple, one-dimensional, dynamic model applied in nontidal rivers, and is particularly used for large rivers.

3. Pollutant Types

   The program can simulate water quality variables such as nitrate, dissolved oxygen, biochemical oxygen demand, ammonium ion, temperature, pH, and other water quality variables.

4. Model Components, Techniques, and Processes

   The model solves ordinary differential equations that are one-dimensional and time-dependent, describing the changes in flows and concentrations. The system is modeled as a series of reaches or segments.

5. Input Data Requirements

   The models assume continuously stirred tank reactors in series. The model requires inputs such as flows from tributaries, points, and diffuse sources. The model also requires flow abstractions.

6. Simulation Outputs

   Output datasets can be used along with the available post-processing tools. Because the output data are obtained in a numerical form, tools such as spreadsheets can be used to process the output.

7. Model Limitations

   In general, the user needs to be familiar with the program architecture, data needs, data output, and so on. The program is user friendly. The outputs can be formatted to graphs and tables.

8. Recommended Reading

Gao, L., & Li, D. (2014). A review of hydrological/water-quality models. *Frontiers of Agricultural Science and Engineering, 1*(4), 267–276.

Lees, M. J., Camacho, L., & Whitehead, P. (1998). Extension of the QUASAR river water quality model to incorporate dead-zone mixing. *Hydrology and Earth System Sciences Discussions, European Geosciences Union,* 2(2/3), 353–365.

Yuceer, M., & Coskun, M. A. (2016). Modeling water quality in rivers: A case study of Beylerderesi river in Turkey. *Inonu University Department of Chemical Engineering, Faculty of Engineering, Inonu University*. Retrieved from https://pdfs.semanticscholar.org/8dee/d2323f37fde813bd0c7e5cf50b743fb9d75c.pdf

3.13 Curvilinear-grid Hydrodynamics 3D Model (CH3D)

1. Availability of the Model

   This model was developed by a researcher currently at the University of Florida. According to the Advanced Coastal Environmental Simulations Laboratory at the University of Florida (https://aces.coastal.ufl.edu/CH3D/), the software is in the public domain, but not necessarily available. Contact information is given on the website for inquiries. Support is also limited.

2. Types of Modeling and Potential Application Areas

   The CH3D is a three-dimensional hydrodynamic model that is suitable for applications to coastal and nearshore waters with complex shoreline and bathymetry.

3. Pollutant Types

   The program can simulate sediment and general water quality such as nitrogen cycle, phosphorus cycle, dissolved oxygen, phytoplankton, and zooplankton.

4. Model Components, Techniques, and Processes

   The model solves three-dimensional, time-dependent mathematical equations for water level, velocities, temperature, and salinity.

5. Input Data Requirements

   The model requires input data such as hydrodynamic information, dimension of the waterbody, and vegetation information.

6. Simulation Outputs

   Output datasets can be used along with the post-processing tools available within the program. Some of the post-processing tools are under development.

7. Model Limitations

   The program requires high computational power to provide appropriate solutions.

8. Recommended Reading

Advanced Coastal Environmental Simulations. CH3D, CH3D-IMS, Next Generation CH3D and CH3D-IMS. *ACES University of Florida*. Retrieved from https://aces.coastal.ufl.edu/CH3D/

3.14 EUTROMOD

1. Availability of the Model

   Originally developed by NALMS, this spreadsheet model is no longer readily available online.

2. Types of Modeling and Potential Application Areas

   The EUTROMOD is a spreadsheet tool that is used for lakes and reservoirs to assess the impact of watershed-derived loads.

3. Pollutant Types

   The model can simulate nitrogen, phosphorus, and algae.

4. Model Components, Techniques, and Processes

   The model uses statistical relationships to determine phosphorus and nitrogen loadings to a lake. The lake response is also derived from regional datasets by establishing statistical relationships.

5. Simulation Outputs

   The tabular data obtained can be used for graphical visualization.

6. Model Limitations

   The program can be used for planning-level estimates. The model uses simplistic relationships to model water quality.

7. Recommended Reading

Ward, G. H., Jr., & Benaman, J. (1999). Models for TMDL application in Texas watercourses: Screening and model review. (CRWR-99-7). Austin, TX: Center for Research in Water Resources at the University of Texas at Austin. Retrieved from https://repositories.lib.utexas.edu/bitstream/handle/2152/6797/crwr_onlinereport99-7.pdf?sequence=2

3.15 Visual Plumes

1. Availability of the Model

   Visual Plumes Version 1.0 (released 2001) is the current version. This model is available without cost from the U.S. EPA's Environmental Modeling Community of Practice website (https://www.epa.gov/ceam/visual-plumes), but it is not supported for systems beyond Windows XP. A summary document and user manual are available, but technical support is not provided.

2. Types of Modeling and Potential Application Areas

   Visual Plumes is a dilution model for assessing the impact from plumes in receiving waters.

3. Pollutant Types

   The model can simulate fate and transport of bacteria and tidal build-up capabilities.

4. Model Components, Techniques, and Processes

   The program has five tabs: diffuser tab, ambient tab, special settings tab, text output tab, and graphical output tab.

5. Input Data Requirements

   The program requires data such as the time series data of flow from the diffuser and information on ambient conditions at various depths.

6. Simulation Outputs

   The program has both the visual and tabular outputs for the user to assess dilution.

7. Model Limitations

   Care should be taken while setting up the depth table.

8. Recommended Reading

Frick, W. E., Roberts, P. J. W., Davis, L. R., Keyes, J., Baumgartner, D. J., & George, K. P. (2003). *Dilution models for effluent discharges* (4th edition). Washington, D.C.: U.S. Environmental Protection

Agency. Retrieved from https://www.epa.gov/sites/production/files/documents/VP-Manual.pdf

3.16 Delft3D

1. Availability of the Model

   This open source software is available on the website dedicated to the Delft3D Community (https://oss.deltares.nl/web/delft3d). Support is available through an open source group on LinkedIn (https://www.linkedin.com/groups/3745991/profile), managed and facilitated by the Delft team. Delft Software Days is a periodic conference offering training, courses, support, and development of the model.
2. Types of Modeling and Potential Application Areas

   Delft3D is a three-dimensional model that is used to investigate hydrodynamics, sediment transport, and water quality for streams, rivers, estuaries, and coastal environments.
3. Pollutant Types

   The program can simulate general water quality parameters such as temperature, dissolved oxygen, nitrogen, phosphorus, bacteria, and sediment.
4. Model Components, Techniques, and Processes

   The program uses advection and diffusion solvers for water quality. It uses the D-flow flexible mesh engine for hydrodynamic simulations on unstructured grids.
5. Input Data Requirements

   The model requires information such as the initial system conditions, flow boundary conditions, transport boundary conditions, salinity, and temperature.
6. Simulation Outputs

   The program provides numerical output of the constituents.
7. Model Limitations

   The challenge is setting up the model grid, the extent of the model, and the design and resolution of the grid.
8. Recommended Reading

   Deltares. (2019). Delft3D-FLOW Simulation of multi-dimensional hydrodynamic flows and transport phenomena, including sediments User Manual (Version 3.15, draft). Retrieved from https://content.oss.deltares.nl/delft3d/manuals/Delft3D-FLOW_User_Manual.pdf

Deltares. (2019). Delft3D-WAVE Simulation of short-crestedwaves with SWAN User Manual (Version 3.05, draft). Retrieved from https://content.oss.deltares.nl/delft3d/manuals/Delft3D-WAVE_User_Manual.pdf

Delft3D Open Source Community. Retrieved from https://oss.deltares.nl/web/delft3d

3.17 Tidal Prism Model

1. Availability of the Model

   Not much user group information is available online.

2. Types of Modeling and Potential Application Areas

   The Tidal Prism Model is a steady state model and can only be applied to marinas where tidal forces are predominant with oscillating flows. It simulates physical transport using the concept of tidal flushing.

3. Pollutant Types

   The program can be used to model up to 10 water quality constituents including fecal coliform bacteria.

4. Model Components, Techniques, and Processes

   The model simulates pollutants in the water column and in the benthic sediment.

5. Input Data Requirements

   Some of the input datasets for the model are inflows, prism volume, suspended sediment, and so on.

6. Simulation Outputs

   The model provides various water quality outputs such as total carbon, total phosphorus, residence time of water, and residence time of the pollutants.

7. Model Limitations

   The model is only applicable to the waterbodies that experience tidal forces such as estuaries with oscillating flows. The program is applied to smaller estuaries and the estuaries are well mixed.

8. Recommended Reading

   Abdelrhman, M., & Campbell, D. E. *Tidal prism modeling of phytoplankton and nitrogen concentrations in Narragansett Bay and its sub-embayments*. Presented at CERF 2011; 21st Biennial Conference of the Coastal and Estuarine Research Federation; Societies, Estuaries and Coasts: Adapting to Change, Daytona Beach, FL,

November 06–10, 2011. Retrieved from https://cfpub.epa.gov/si/si_public_record_report.cfm?dirEntryId=234783&Lab=NHEERL

4.0 WATER QUALITY COMPLIANCE MODELING

4.1 System for Urban Stormwater Treatment and Analysis Integration (SUSTAIN)

1. Availability of the Model

   The SUSTAIN model was originally developed by U.S. EPA. Version 1.2 is available for download without cost from the SUSTAIN webpage on U.S. EPA's website (https://www.epa.gov/water-research/system-urban-stormwater-treatment-and-analysis-integration-sustain). Although some documentation exists, U.S. EPA no longer develops or supports the SUSTAIN model and has no plans to develop versions that can work with newer versions of GIS or Windows. Innovyze has taken U.S. EPA's model and incorporated it to their InfoSWMM product as InfoSWMM Sustain. Support for this version is available.

2. Types of Modeling and Potential Application Areas

   The program is an environmental decision support system with a GIS interface that can simulate stormwater runoff, water quality components such as sediment, nutrients from one or multiple watersheds, and different LID BMPs to meet water quality objectives. The software helps practitioners to evaluate a combination of LID BMPs based on cost and effectiveness using the cost-optimization tool available within the program. Additionally, the program has a BMP siting tool to locate suitable places for the BMPs based on the site suitability criteria such as drainage area, slope, hydrologic soil group, groundwater table depth, road buffers, stream buffers, and building buffers. The current version of SUSTAIN has the following LID BMPs: porous pavement, green roof, bioretention cell, cistern, rain barrel, infiltration basin, infiltration trench, grassed swale, vegetative filter strip, sand filter, constructed wetland, dry pond, and wet pond. The program can be applied to develop TMDL implementation plans, identifying management strategies to meet pollutant load reduction goals assigned to a municipal separate storm sewer system (MS4), determining strategies to reduce volume and peak flows to combined sewer systems, and evaluating cost benefits of LID BMPs. The tool has been used in watersheds in the Los Angeles area to evaluate the capabilities of the BMPs to meet water quality goals.

3. Pollutant Types

   The program can simulate runoff, sediment, and nutrients.

4. Model Components, Techniques, and Processes

   The program can run internal land simulation using the algorithms adapted from SWMM to generate hydrograph and pollutograph and also has an external land simulation option where the program receives the time series of flow and pollutant information in hourly time step for each land use from a precalibrated model such as HSPF/LSPC. Stage-outflow storage routing is used for flow routing. Green and Ampt, Holtan–Lopez, and Horton infiltration methods are available in the program. Sediment algorithms have been adopted from HSPF.

5. Input Data Requirements

   This GIS-based program requires watershed loading, BMP characteristics, and treatment train configuration.

6. Simulation Outputs

   The model provides tabular and graphical outputs for flow and pollutants. Microsoft Excel is used as the post-processor to analyze and interpret simulation results. The output is obtained in hourly or subhourly time steps. The post-processor can be used to visualize and manage output to assess treatment capabilities, cost-effectiveness curve, BMP selection, and overall costs, among other parameters.

7. Model Limitations

   The program cannot perform instream water quality simulations. Currently, the model is not supported by U.S. EPA.

8. Recommended Reading

Shoemaker, S., Riverson, J., Jr., Alvi, K., Zhen, J. X., Paul, S., & Rafi, T. (2009). *SUSTAIN - A framework for placement of best management practices in urban watersheds to protect water quality* (EPA/600/R-09/095). U.S. EPA. Retrieved from https://www.epa.gov/sites/production/files/2015-10/documents/sustain_complex_tools.pdf

U.S. EPA. (n.d.). System for Urban Stormwater Treatment and Analysis Integration (SUSTAIN). U.S. EPA. Retrieved from https://www.epa.gov/water-research/system-urban-stormwater-treatment-and-analysis-integration-sustain

4.2 Best Management Practice Treatment options for Removal on an Annual Basis by Those Interested in Nutrients in Stormwater (BMPTRAINS)

1. Availability of the Model

   Version 8.6 is the most current model (released 2018), and it is available for download free of charge from the University of Central Florida (https://stars.library.ucf.edu/bmptrains/). A user manual, video tutorials, and numerous publications are available on their website.

2. Types of Modeling and Potential Application Areas

   The spreadsheet-based computer program is used to assess the impacts of stormwater using BMPs.

3. Pollutant Types

   The program estimates removal of nitrogen and phosphorus.

4. Model Components, Techniques, and Processes

   The model has an estimation of the required annual removal efficiency of the pollutants from the BMP and annual loadings from the development and estimation of annual runoff volumes, which is estimated based on meteorological zone location, watershed area, mean rainfall, directly connected impervious areas (DCIA) percentages and non-DCIA.

5. Input Data Requirements

   The model requires watershed characteristics, meteorological information, and BMP information.

6. Simulation Outputs

   The model provides a summary output report.

7. Model Limitations

   The model is a spreadsheet-based model.

8. Recommended Reading

   University of Central Florida. BMPTRAINS home page. Retrieved from https://stars.library.ucf.edu/bmptrains/

   Wanielista, M. (2019). User Manual—BMP Trains 2020 Newest Version. *BMP Trains*. 25. Retrieved from https://stars.library.ucf.edu/bmptrains/25/

4.3 Virginia Runoff Reduction Method (VRRM)

1. Availability of the Model

   This spreadsheet is available free of charge from the Virginia Department of Environmental Quality (DEQ) on the Stormwater BMP

Clearinghouse website hosted by Virginia Tech (https://www.deq.virginia.gov/Programs/Water/LawsRegulationsGuidance/Guidance/StormwaterManagementGuidance.aspx). Version 3.0 is the most recent version. A guidance document and limited support are available from DEQ.

2. Types of Modeling and Potential Application Areas

   The VRRM is a spreadsheet-based program that is designed to help planners to meet stormwater standards within the State of Virginia.

3. Pollutant Types

   In general, the program models total phosphorus and total nitrogen.

4. Model Components, Techniques, and Processes

   The program estimates annual post-development nutrient loadings and enables the user in the selection of the BMPs to reduce the loading and meet the regulatory standard.

5. Input Data Requirements

   The program requires site information including rainfall, watershed area, land cover, and information on runoff reduction practices.

6. Simulation Outputs

   The program displays results in the water quality compliance worksheet.

7. Model Limitations

   The model is a spreadsheet model and has straightforward water quality calculations.

8. Recommended Reading

   Commonwealth of Virginia Department of Environmental Quality. (2016). *Virginia Runoff Reduction Method compliance spreadsheet user's guide & documentation version 3.0* (No. 16-2001). Commonwealth of Virginia DEQ. Retrieved from https://www.deq.virginia.gov/Portals/0/DEQ/Water/StormwaterManagement/VRRM/GM14-2001%20Virginia%20Runoff%20Reduction%20Method_V3.pdf

4.4 Integrated Design, Evaluation, and Assessment of Loadings (IDEAL)

1. Availability of the Model

   The IDEAL model was developed by Woolpert, Inc. in collaboration with Dr. Bill Barfield of Oklahoma State University and Dr. John Hayes of Clemson University. IDEAL has been distributed and technically supported under the trade name StormOps since 2006 (https://www.stormopssoftware.com/). The Greenville County, South Carolina, version is available free of charge to Greenville County residents; versions tailored to other locations are available for a fee. A user manual is included with the software, and support is included with the initial software purchase (and may be renewed with an annual licensing fee). The most recent version is 3.0.0 (released 2019).

2. Types of Modeling and Potential Application Areas

   The IDEAL model is capable of simulating runoff, total suspended solids, nutrients, and bacteria loadings from single storms and annual storms on a site level to meet regulatory standards. In general, the program is used in assessing the effectiveness of post-construction BMPs in urban and semi-urban settings. Users can model single BMP or multiple BMPs. Currently, the user can model dry detention ponds, wet detention ponds, vegetative filter strips, bioswales, enhanced bioswales, infriltration trenches, porous pavements, sand filters, bioretention cells, cisterns, and user-defined BMPs. The model is used extensively by Greenville County, South Carolina, for post-construction permit compliance. The tool has also been used by other MS4s in South Carolina for assessing water quality compliance and TMDL needs.

3. Pollutant Types

   The program can simulate runoff, total suspended solids, total phosphorus, total nitrogen, and bacteria.

4. Model Components, Techniques, and Processes

   The program can simulate watershed runoff from single storm and annual storms using the NRCS curve number method. Simple translation or the Muskingum-Cunge method is used for flow routing. Infiltration in infiltration BMPs is governed by Green and Ampt and Darcy's model. The modified universal soil loss equation (MUSLE) is used for sediment yield from pervious areas, and event mean concentration (EMC) is used for impervious areas. Nutrient and bacteria yield are computed using EMC and runoff volume.

5. Input Data Requirements

   Watershed inputs include size, soil, slope, and land use. The BMP inputs include dimensions, riser configuration, if applicable, and other characteristics.

6. Simulation Outputs

   The model provides numerical outputs for flow, sediment, nutrient, and bacteria at the end of the simulation in form of a report. It also provides graphical output for each object (e.g., watershed, BMPs).

7. Model Limitations

   The program has not been implemented in large-scale watershed simulations. It does not have instream water quality simulation capabilities. The program is unable to handle downstream disturbances such as flow restriction and tailwater. The model should not be used to design conveyances.

8. Recommended Reading

   Barfield, B. J., Hayes, J. C., Holbrook, K. F., Bates, B., Gillespie, J. & Fersner, J. (2003). IDEAL: A model of runoff, TSS, and nutrient yield from small urban watersheds. Proceedings of the 2003 EWRI Annual Meeting, Philadelphia, Pennsylvania, June 23–26. New York, NY: American Society of Civil Engineers.

   Hayes, J. C., Barfield, B., Holbrook, K. F., Gillespie, J., Fersner, J. & Bates, B. (2006). A model for assessing the impact of BMPs on water quality. *Stormwater Magazine*. Retrieved from http://media.wix.com/ugd/e6190c_bd486891290343dc8e64a3c1646a117e.pdf

   Stormwater Quality Modeling Software. Retrieved from https://stormopssoftware.com/

5.0 INTEGRATED MODELING SYSTEMS

5.1 Better Assessment Science Integrating Point and Non-Point Sources (BASINS)

1. Availability of the Model

   This model's current version, 4.5, is available for download without cost from U.S. EPA's Environmental Modeling Community of Practice (https://www.epa.gov/ceam/better-assessment-science-integrating-point-and-non-point-sources-basins). A listserv is available for modeling assistance, as well as an email address for additional technical questions. Documentation, user manual, tutorials, training information, case studies, and technical notes are all available on the U.S. EPA website.

2. Types of Modeling and Potential Application Areas

   The BASINS program is an interface to develop input files for different models and assist in communication among various computer programs. It is an integrated environmental analysis system designed to assist in watershed studies. The program works within a GIS environment, which helps in managing spatial information. In addition to generating input files for the models, BASINS also assists in executing the watershed and water quality models such as SWAT, HSPF, WASP, PLOAD, and SWMM. The program is extensively used in several parts of the country. Given the complexity of each model, the BASINS interface has made the modeling effort easier by shrinking down the time-consuming modeling efforts.

3. Pollutant Types

   The watershed water quality models that come with BASINS can simulate runoff, sediment, nutrients, and other pollutants.

4. Model Components, Techniques, and Processes

   Because BASINS is a modeling platform, each program simulates different hydrologic components and fate and transport of pollutants using established science. Those simulation techniques may or may not vary across different programs. For detailed information on the individual model, please refer to the fact sheet of the respective model.

5. Input Data Requirements

   A wide range of data needs may be required depending on the desire level of detail and systems being modeled. Generally, much of the input data for BASINS models (watershed, meteorological, soils, and land use) are available in the program itself. The user will need to supply point-source loads and some other watershed characteristics.

6. Simulation Outputs

   The BASINS model has tools to read the outputs from the model and display the output in a graphical form.

7. Model Limitations

   The program works on a GIS platform, which could limit a non-GIS user.

8. Recommended Reading

   U.S. EPA. (n.d.). Better Assessment Science Integrating Point and Non-Point Sources (BASINS): Introduction. U.S. EPA. Retrieved from https://www.epa.gov/ceam/better-assessment-science-integrating-point-and-non-point-sources-basins

5.2 Watershed Modeling Systems (WMS)

1. Availability of the Model

   This model was developed by the Environmental Modeling Research Laboratory of Brigham Young University in conjunction with the USACE Waterways Experiment Station. The program is supported by Aquaveo, LLC. Several "levels" of the current version, 11.0, are sold on their website (https://www.aquaveo.com/software/wms-watershed-modeling-system-introduction). The basic version is free. Tutorials, videos, documentation, training courses, on-site training, and an online community forum are also available through their website.

2. Types of Modeling and Potential Application Areas

   The program is applied to understand various watershed hydrologic and hydraulic conditions.

3. Pollutant Types

   The program can assist in modeling the general pollutants such as temperature, dissolved oxygen, nitrogen, phosphorus, sediment, bacteria, pesticide, metals, and algae.

4. Model Components, Techniques, and Processes

   The program provides various tools to create inputs for the modeling purposes. The program supports various models such as HEC-1, HEC-RAS, HEC-HMS, Natural Resources Conservation Service Technical Release 20 (TR-20), Technical Release 55 (TR-55), National Flood Frequency (NFF), MODRAT, and CE-QUAL-W2.

5. Input Data Requirements

   Depending on the model, the program generates all the input data using a GIS framework. The user of the program creates all the input files pertinent to the model.

6. Simulation Outputs

   The model outputs in numerical format and has the interface for graphical display.

7. Model Limitations

   The program is not a watershed computer model by itself. This is a program supported by Aquaveo, LLC.

8. Recommended Reading

Aquaveo, LLC. (n.d.). WMS—The All-in-one Watershed Solution. Aquaveo, LLC. Retrieved from https://www.aquaveo.com/software/wms-watershed-modeling-system-introduction

WMS: Watershed Modeling System. (n.d.). University of California, Davis. Retrieved from https://cwam.ucdavis.edu/pdfs/wms.pdf

Index

CPSIA information can be obtained
at www.ICGtesting.com
Printed in the USA
FSHW020142230220
67300FS

9 781572 783591